The Explanatory Autonomy of the Biological Sciences

This book argues for the explanatory autonomy of the biological sciences. It does so by showing that scientific explanations in the biological sciences cannot be reduced to explanations in the fundamental sciences such as physics and chemistry and by demonstrating that biological explanations are advanced by models rather than laws of nature.

To maintain the explanatory autonomy of the biological sciences, the author argues against explanatory reductionism and shows that explanation in the biological sciences can be achieved without reduction. Then, he demonstrates that the biological sciences do not have laws of nature. Instead of laws, he suggests that biological models usually do the explanatory work. To understand how a biological model can explain phenomena in the world, the author proposes an inferential account of model explanation. The basic idea of this account is that, for a model to be explanatory, it must answer two kinds of questions: counterfactual-dependence questions that concern the model itself and hypothetical questions that concern the relationship between the model and its target system. The reason a biological model can answer these two kinds of questions is due to the fact that a model is a structure, and the holistic relationship between the model and its target warrants the hypothetical inference from the model to its target and thus helps answer the second kind of question.

The Explanatory Autonomy of the Biological Sciences will be of interest to researchers and advanced students working in philosophy of science, philosophy of biology and metaphysics.

Wei Fang is Associate Professor in the Research Centre for Philosophy of Science and Technology, Shanxi University, China. His research topics include scientific explanation, models and modeling, causal modeling, and mechanisms, among others, and has published a number of papers in, among others, *Philosophy of Science*, *Biology & Philosophy*, and *Synthese*.

Routledge Studies in the Philosophy of Science

The Instrument of Science
Scientific Anti-Realism Revitalised
Darrell P. Rowbottom

Understanding Perspectivism
Scientific Challenges and Methodological Prospects
Edited by Michela Massimi and Casey D. McCoy

The Aesthetics of Science
Beauty, Imagination and Understanding
Edited by Steven French and Milena Ivanova

Structure, Evidence, and Heuristic
Evolutionary Biology, Economics, and the Philosophy of Their Relationship
Armin W. Schulz

Scientific Challenges to Common Sense Philosophy
Edited by Rik Peels, Jeroen de Ridder, and René van Woudenberg

Science, Freedom, Democracy
Edited by Péter Hartl and Adam Tamas Tuboly

Logical Empiricism and the Physical Sciences
From Philosophy of Nature to Philosophy of Physics
Edited by Sebastian Lutz and Adam Tamas Tuboly

The Explanatory Autonomy of the Biological Sciences
Wei Fang

For more information about this series, please visit: https://www.routledge.com/Routledge-Studies-in-the-Philosophy-of-Science/book-series/POS

The Explanatory Autonomy of the Biological Sciences

Wei Fang

NEW YORK AND LONDON

First published 2022
by Routledge
605 Third Avenue, New York, NY 10158

and by Routledge
2 Park Square, Milton Park, Abingdon, Oxon, OX14 4RN

Routledge is an imprint of the Taylor & Francis Group, an informa business

Library of Congress Cataloging-in-Publication Data
Names: Fang, Wei (Associate Professor), author.
Title: The explanatory autonomy of the biological sciences / Wei Fang.
Description: New York : Taylor & Francis, 2021. | Series: Routledge research in aesthetics | Includes bibliographical references and index.
Identifiers: LCCN 2021032607 (print) | LCCN 2021032608 (ebook) | ISBN 9780367693510 (hardback) | ISBN 9781003148029 (ebook)
Subjects: LCSH: Biology--Philosophy. | Autonomy (Philosophy)
Classification: LCC QH331 .F25 2021 (print) | LCC QH331 (ebook) | DDC 570.1--dc23
LC record available at https://lccn.loc.gov/2021032607
LC ebook record available at https://lccn.loc.gov/2021032608

ISBN: 978-0-367-69351-0 (hbk)
ISBN: 978-0-367-70788-0 (pbk)
ISBN: 978-1-003-14802-9 (ebk)

DOI: 10.4324/9781003148029

Typeset in Sabon
by SPi Technologies India Pvt Ltd (Straive)

Printed in the United Kingdom
by Henry Ling Limited

To my most important mentors,
Paul and Arnaud

Contents

Acknowledgments

This book is based on my doctoral dissertation "The Explanatory Autonomy of the Biological Sciences", which was submitted to the University of Sydney Philosophy Department in January 2017. My dissertation was supervised by Paul Griffiths (Sydney) and Arnaud Pocheville (Centre national de la recherche scientifique), and I would like to express my sincere gratitude to both of them for their support, patience and encouragement during the years I spent at the University of Sydney. Some chapters, with varying degrees of modifications, have already been published; for example, Chapter 2 on multiple realization first appeared as Fang (2018), Chapter 5 on models and modeling first appeared as Fang (2017) and Chapter 7 on model explanation first appeared as Fang (2019b).

My way of answering the question, "How the biological sciences can be explanatorily autonomous?" was significantly influenced and shaped by the members of the Theory and Method in Biosciences group at the Charles Perkins Centre, University of Sydney, a group led by Paul Griffiths and whose members were not only philosophers of biology but also biologists. Scholars in this group took a "philosophy in science" approach to investigating philosophical problems, an approach I adopted in my own writing and thinking. I am especially grateful to Pierrick Bourrat, who helped me in numerous ways during the past a few years, and some of my ideas were inspired by discussions with him. Other group members also played an indispensable role in shaping my ideas in this book, including Adam Hochman, Brett Calcott, Karl Rollings, Elena Walsh, Isobel Ronai, John Matthewson, Karola Stotz, Kate Lynch, Michael Levot, Qiaoying Lu, Stefan Gawronski and Xin Zhang.

Special thanks are due to Patrick McGivern, Jay Odenbaugh and Wendy Parker, who were the examiners of my original PhD thesis and offered a great many valuable comments and suggestions for improving the thesis. I would also like to thank many colleagues in this field who have provided me with immense help: Brian Hedden, David Braddon-Mitchell, Jing Wang, Kristie Miller, Mark Colyvan, Maureen O'Malley, Michael Weisberg, Otávio Bueno, Patrick McGivern, Peter Godfrey-Smith,

Stephen Downes, Xiang Huang, and Zhilin Zhang. Also, a special thanks to Maureen O'Malley, who introduced me to the world of the philosophy of models and modeling. Her interesting and stimulating course on the philosophy of models and modeling at the University of Sydney was the starting place from which my interest in models and modeling was built. I also want to say thanks to two anonymous referees and the editors of this book series for their helpful comments on earlier drafts of the book.

Finally, my sincere thanks also go to my family. Words cannot express how grateful I am to my mother, father and little sister, who always believe in me and encourage me to follow my dreams.

1 Introduction

The dispute surrounding the autonomy of the biological sciences has a long history, which can be traced back to more than half a century ago (e.g., Mainx 1955; Smart 1963; Simpson 1964; Ayala 1968; Mayr 1969; Ruse 1973; Munson 1975). The dispute stemmed from the concern about the position of biology in the natural sciences, and one prominent view back in the 1950s and 1960s was that biology is a province of physics, which can finally be reduced to physics. For instance, Francis Crick, a father of modern genetics, once claimed that "the ultimate aim [of science] … is in fact to explain all biology in terms of physics and chemistry" (Crick 1966, 10; cf. Mayr 1996, 98). However, defenders argued that "biology fully merits status as an autonomous science because it differs fundamentally in its subject matter, conceptual framework, and methodology from the physical sciences" (Ayala 1968; cf. Mayr 1988, 8). Since then defending or refuting the autonomy of biology has become a central problem in the philosophy of biology, as Rosenberg said in the 1980s that "whether and how biology differs from the other natural sciences … is the most prominent, obvious, frequently posed, and controversial issue the philosophy of biology faces" (Rosenberg 1985, 13; cf. Mayr 1996, 98).

From the late 1960s to the 1990s, the dispute largely shifted to the question of whether theories in one domain can be reduced to another domain within biology. In particular, the main focus was on whether and in what sense classical genetics can be reduced to molecular biology (e.g., Schaffner 1967, 1969, 1974, 1976, 1993, 1996; Ruse 1971, 1976; Hull 1972, 1974, 1976, 1979; Wimsatt 1974, 1976, 1979; Kitcher 1984; Rosenberg 1985, 1997; Sarkar 1992, 1998). More recently, due to shortcomings of the theory reductionist approach, an explanatory reductionist alternative has been proposed by several major figures in the philosophy of biology, such as Waters (1990, 1994, 2000, 2008), Sarkar (2001, 2002, 2005), Weber (2005), and Schaffner (2006), among others. While the theory reductionist view requires reducing one whole theory to another, the explanatory reductionist approach merely requires that the feature represented in a higher-level science (e.g., ecology) can be reductively explained in virtue of features represented in a lower-level science (e.g.,

DOI: 10.4324/9781003148029-1

molecular biology) or that elements of a higher-level science (e.g., generalizations, descriptions, laws, models, etc.) can be reductively explained by elements of a lower-level science.

This book concentrates on the explanatory autonomy side of the biological sciences, which concerns problems such as explanatory reductionism, laws of nature, biological models, and so on. By the term *the biological sciences*, I mainly refer to areas of biology other than molecular biology, such as developmental biology, cell biology, physiology, ecology, and neuroscience, to list just a few. Hence, to be more precise, the focus of this book is on the explanatory autonomy of the biological sciences other than molecular biology. Alternatively, if we were concerned with the explanatory autonomy of molecular biology, then the focus would be on the relationship between molecular biology and chemistry (or biochemistry or, more broadly, the physical sciences). Narrowly understood in this way, the explanatory autonomy problem sometimes comes down to the relationship between different areas of the biological sciences (excluding molecular biology) and sometimes comes down to the relationship between the various biological sciences and molecular biology. As a result, this book does not consider the traditional problem concerning the relationship between biology and physics (or chemistry).

The book has two interrelated parts. The first part attempts to argue against one influential form of reductionism, that is, epistemic or explanatory reductionism, while the second part proposes a positive account of how explanation in the biological sciences is possible without reduction (and without laws). The idea behind the positive account is that, although the biological sciences do not have laws and do not borrow their explanatory power from other domains by reduction, they are explanatorily autonomous because they usually explain using biological models.

Part I discusses two important forms of reductionism: metaphysical and epistemic/explanatory reductionism.[1] The reason for discussing metaphysical reductionism in a book defending the explanatory (epistemic) autonomy of the biological sciences is simply that the former may have some implications for the latter. The layout of this part is as follows. Chapter 2 deals with the relationship between properties (or relations, processes, states of affairs, etc.) at different levels of organization, for example, molecular versus cellular (Longo et al. 2012). One starting point for discussing this relationship is the multiple realization (MR) thesis, a thesis that states roughly that different lower-level properties can realize the same higher-level property. For example, one may say that both cephalopods and human beings feel pain, but the underlying physical processes involved are substantially different: they have the same mental property but with different physical structures.

Yet there is a substantial debate surrounding the plausibility of the MR thesis. Among many critics of the thesis, Shapiro's *causally relevant differences* criterion is so influential that it recently has resuscitated the debate (Shapiro 2000). Much clarification and refinement have occurred since the somewhat abstract criterion was first proposed (Shapiro 2008; Polger 2009; Shapiro and Polger 2012). In a very recent work, the criterion has been most precisely articulated by Polger and Shapiro (2016) as the Official Recipe for MR.

With Polger and Shapiro's criterion at hand, we are in a good position to evaluate when a case of MR is present. My disagreement with Polger and Shapiro's criterion concerns only the scope, rather than the substance, of MR. Although Polger and Shapiro claim that their criterion makes MR less common than its proponents envision, I argue that their criterion makes MR less rare than they suggest. To show this, I first develop a mechanistic multiple realization thesis (MMRT). Then, two would-be cases of MR are examined in terms of the MMRT, showing that MR is prevalent in the biological world. Finally, I consider the potential implications of the MMRT for the reductionism versus antireductionism debate in the philosophy of biology. It turns out that, on its own, the MMRT does not favor either side of the reductionism debate. In particular, the MMRT neither supports nor counts against the explanatory reductionist position (which is a version of epistemic reductionism and the focus of this book). I suggest that, to take sides in the debate, the MMRT must be allied with an account of scientific explanation, which is discussed in Chapter 3.

Chapter 3 argues against explanatory reductionism. The basic idea of this form of reductionism is summarized as a higher-level phenomenon/explanation that can be further explained by a relatively lower-level phenomenon/explanation, and the explanatory significance is primarily, if not completely, located at the lower level. In philosophy of biology, the lower level is usually referred to as the molecular level. To argue against this form of reductionism, I consider three lines of argument. I first consider the context argument proposed by antireductionists. There are two ways to raise context arguments. The first is the context-dependence argument, which states that, depending on different contextual factors (cellular, organelles, tissues, organismic, or even environmental), a molecular entity can give rise to many and sometimes even radically distinct effects. The second is the context-independence argument, which states that higher-level phenomena can be independent of their lower-level underpinnings, because changing the lower-level underpinnings from one to another does not necessarily result in changes to the higher-level phenomena. The context-independence argument is reminiscent of the multiple realization thesis discussed in Chapter 3. The claim is that the MR thesis can be employed in the context of scientific explanation as a foundation for arguing against reductionism, because the thesis states that there are many different lower-level underpinnings that realize the same higher-level phenomenon. That is, the MR thesis guarantees that the higher-level phenomenon can be partially independent of its lower-level underpinnings because changing the lower-level underpinnings from one to another does not necessarily result in changes in the higher-level phenomenon. In the context of scientific explanation, this means that explanation can proceed at a level of analysis without making an appeal to its underlying happenings.

The two arguments considered together show that, in the arena of scientific explanation, the strategy of holding certain elements fixed as contexts runs both ways: higher-level elements are routinely treated as context when exploring lower-level elements, and lower-level elements are usually treated as context when exploring higher-level elements. In either case, a one-size-fits-all explanatory reductionist scenario does not arise. For the former, it is not a *bona fide* reductionist strategy, because contextual elements that may be causally relevant to generating a phenomenon of interest are not fully specified or reduced, nor can they be fully specified or reduced. For the latter, it is not a genuine reductionist strategy, because the explanation usually operates at a level of analysis relatively independent of its underlying molecular foundation.

A second line of argument against explanatory reductionism is based on the fact that explanations in many areas of biology (e.g., developmental biology, ecology, systems biology, etc.) cannot be achieved without additional knowledge obtained from domains other than molecular biology. This kind of knowledge may include, but is not restricted to, spatial, structural, geometrical, or topological information that cannot be simply subsumed into, or reduced to, the domain of molecular biology. Geometrical explanation is one case in point. To show that geometrical information plays a central role in explaining some biological phenomena, an example drawn from Philippe Huneman (2010) is examined. The example shows that the relevant geometrical properties are properties of the whole system, relatively independent of their underlying mechanistic underpinnings. The general lesson is that, although geometrical explanation may not fully replace reduction-oriented mechanistic explanation, it is clearly not the case that scientists can explain certain biological phenomena by appealing exclusively to molecular information.

The third line of argument against explanatory reductionism highlights the fact that scientific explanation is, though ontic-oriented, inherently pragmatically driven. The central claim is that there is no better explanation *simpliciter*, because explanations are characteristically advanced for the purpose of answering specific questions. Therefore, the notion of a superior explanation can only be legitimately understood relative to specific questions being asked in specific contexts.

The first part concludes by making the following points: First, even if some version of the multiple realization thesis can be established (such as the multiple mechanistic realization thesis), it is, by itself, neutral with respect to both metaphysical and epistemic reductionism. Second, to argue against epistemic reductionism, the MR thesis must be allied with a conception of scientific explanation. In addition, the first part provides an ontic-oriented epistemic conception of scientific explanation that highlights the following three elements: contextual elements, extra information, and pragmatic considerations. Although this list of dimensions is far from exhaustive, it is sufficient to show that explanations in the biological sciences do not always operate in a reductionist fashion.

The second part begins with Chapter 4, where we face the problem of how the biological sciences can be explanatory if they do not have laws of nature. Part of the motivation for holding a reductionist stance in the biological sciences originates from the worry that only laws can explain—this is problematic if the biological sciences do not have laws. Hence, the suggested solution to the absence of laws is that, if the biological sciences can be reduced to the *hard sciences* (e.g., chemistry or physics) that do have laws, then their explanatory power is warranted. With respect to the problem of laws in the biological sciences, there seem to be three options: (a) the biological sciences do have laws so they can be explanatory, (b) the biological sciences themselves do not have laws but they are based on certain underlying physicochemical laws (from which their explanatory power is derived), and (c) if the biological sciences do not have laws and are not based on physicochemical laws, then philosophers must provide an account of how they can be so explanatory—since only laws can do explanatory work (Hempel 1965). The first option is held by Sober (1984, 1997, 2000, 2008, 2010), Lange (1993, 2002, 2005), Mitchell (1997, 2000), and Elgin (2003, 2006); the second by Rosenberg (2001a, 2001b, 2006, 2012) and Weber (2005); and the third by Rosenberg (2001c).

In Chapter 4, I explain why I disagree with all these options. In response to the first option, I argue that there are no laws in the biological sciences. In response to the third option, I claim that philosophically adequate accounts of explanation are available that do not invoke the notion of laws of nature. I do not discuss the second option in this chapter. This is because, on the one hand, I have shown in the first response that there are no laws in the biological sciences, and, on the other, I have already argued in Chapter 3 that scientific explanation is inherently a pragmatic matter wherein lower-level explanation (e.g., a physicochemical explanation) is not necessarily better than higher-level explanation. It follows that the second option is not promising.

One may worry that the absence of laws renders the explanatory power of the biological sciences mysterious. Yet I suggest that explanatory power can be shouldered by another kind of explanatory vehicle: biological models. The corresponding claim is that the biological sciences are abundant in models rather than laws. That is, making an appeal to laws is unnecessary because it is models rather than laws that play the central explanatory role in the biological sciences. To set the stage for the discussion of the following chapters, two models are introduced: the leaf gas-exchange model and the San Francisco Bay model. To be clear, the

Bay model is not a biological model. Nevertheless, introducing a nonbiological example shows that, although my view on models and modeling is built on investigating biological models, it can be further generalized to nonbiological models.

Chapter 4 points to a way to maintain the explanatory autonomy of the biological sciences. Using models, we are in a good position to understand why the biological sciences can be explanatory. My understanding of how biological models can be explanatory proceeds in two steps. The first step is to understand what relationship the model bears to its target system, namely, the model–world relationship. The second step is to use the model–world relationship as a basis for the development of an account of how models can be explanatory. A good starting point to fulfill the first step is to look at one very influential account in the literature, namely, Michael Weisberg's weighted feature-matching account (Weisberg, 2013), a version of the similarity view. This is done in Chapter 5.

Weisberg's account concentrates on the ways in which models are similar to their target systems. He intends not only to explain what similarity consists in but also to capture similarity judgments made by scientists. In order to scrutinize whether his account accomplishes this goal, I outline one common way by which scientists judge whether a model is similar enough to its target, namely, the maximum likelihood estimation (MLE) method. The leaf gas-exchange model is examined using this estimation method. I then consider whether Weisberg's account can capture the judgments involved in MLE practice. I show that Weisberg's account fails for three reasons. First, his account is simply too abstract to capture what is going on in the MLE. Second, it implies an atomistic conception of similarity, while the MLE operates in a holistic manner. Third, Weisberg's atomistic conception of similarity can be traced back to a problematic set-theoretic approach to the structure of models.

The failure of Weisberg's account calls for an alternative. Chapter 6 endeavors to propose such an alternative. In particular, through looking closely at biological modeling practice, a holistic view of the model–world relationship is proposed. The holistic view holds that the model–world relationship constitutes a holistic fit, where *holistic fit* refers to the degree to which the model has the same structure as its target system or refers to the distance between two structures. The MLE practice is only roughly described in Chapter 5 for the purpose of unveiling the shortcomings of Weisberg's account. In this chapter, it is reexamined in depth with the hope of deriving the philosophical implications pertaining to fleshing out the holistic view. To pave the way for reexamining the MLE practice, I first discuss a simpler estimation method also commonly used in practice, namely, the least squares estimation (LSE) method, and then come back to the MLE practice. The lesson learnt through a careful examination of these two methods is that, in practice, modelers always compare the model and its target system holistically.

After this discussion, I tentatively suggest that, although part of the goal of this chapter is to develop a holistic view about biological models, the philosophical implications derived from discussing biological models seem to be very general such that they can be extrapolated to nonbiological models. To this end, a concrete model, that is, the San Francisco Bay model, is discussed. Given that the notion of structure is also employed in the semantic view of models (Suppes 1960, 1962; Sneed 1971; Suppe 1977, 1989; Stegmüller 1976; van Fraassen 1980; Thompson 1983; Lloyd 1994), I feel the need to say more about what I mean by the term *structure*. For this purpose, a deflationary account of model structures is also proposed in this chapter. The basic idea is that model structures, of various kinds (e.g., concrete physical structures, equations, graphs, pictures, abstract structures in logic and meta-mathematics, etc.), are important *inferential tools* in modeling practice. Nonetheless, although model structures can be of various kinds, they have one key common feature: they are, or at least can be described as, *dependence relationships*. Furthermore, I claim that genuine dependence relationships (causal or noncausal) are those that allow one to answer Woodward's *what-if-things-had-been-different* questions ("w-questions"). That is, they tell us how certain variables would change if other variables were to be changed or manipulated (Woodward 1997, 2003, 2010; also see Woodward and Hitchcock 2003).

Based on the first step of articulating what the model–world relationship consists of, Chapter 7 attempts to develop a holistic account of how biological models can be explanatory. Inspired in large part by Woodward's interventionist account of scientific explanation and Bokulich's structural account of model explanation, I claim that, for a model to be explanatory, it must help the modelers to answer two kinds of questions: counterfactual dependence questions (corresponding to Woodward's w-questions) that concern the model itself and *hypothetical* questions that concern the relationship between the model and its target system. Furthermore, I suggest that the reason a biological model can answer these two kinds of questions is due to the fact that (a) a model is a structure, that is, a set of dependence relationships that can be employed to answer w-questions, and that (b) the holistic fit between the model and its target warrants the hypothetical inference from the model to its target and thus helps answer the second kind of question.

I call the second kind of relationship a *hypothetical relationship*, both because it features *hypothetical reasoning* in modeling practice,[2] and it also directly relates to Giere's (1988, 85) notion of *theoretical hypothesis* that, similarly to my own notion, links the model to its target in the real world. By and large, hypothetical relationships are useful *heuristics* whereby scientists make inferences (i.e., explanations and predictions) about the world using the model. The hypothetical inference connecting the model and its target often takes the following form: if M is a good model and if *M* has such and such attributes, patterns, or mechanisms (call these "attributes" for short), then, *hypothetically*, the target *T* would also have such and such attributes. The fact that—given *M* is a good model—*M* has such and such attributes can lead to the *prediction* that *T* may also have such and such attributes. On the other hand, exploring the way the model produces such and such attributes leads to the *explanation* of why *T* manifests such and such attributes. Therefore, part of the reason a model can be explanatory (and predictive) is due to the fact that, on the basis of the holistic fit between the model and its target system, the counterfactual structure of the model may be extrapolated *to* its target in terms of the *hypothetical relationships* that have been built between the model and the target.

The second part achieves the following: Chapter 4 shows that there are no laws in the biological sciences and suggests that it is models rather than laws that play the explanatory role in the biological sciences. Second, to pave the way for proposing an account of how biological models can be explanatory, Chapter 5 examines the shortcomings of Weisberg's similarity view of the model-world relationship, while Chapter 6 outlines a holistic alternative to Weisberg's view. Finally, Chapter 7 develops a holistic view of model explanation in the biological sciences.

All in all, the book accomplishes two interrelated tasks: one is that there are far fewer instances of higher-level explanations in the biological sciences that are reducible to lower-level ones. The second is that biological explanations proceed typically through the use of models, without reliance on either reduction or laws of nature. Therefore, the biological sciences do have their own explanatory autonomy.

Notes

1 According to Hüttemann and Love (2011), "explanatory reductionism is only one dimension of epistemic reductionism" (524). Methodological reductionism is another. They are relatively independent theses because neither entails the other, for example, "methodological reductionism does not guarantee explanatory success and a successful explanatory reduction does not imply that methodological reduction is the most favorable strategy of inquiry" (ibid., 524). This book focuses on the dimension of explanatory reductionism. Hence, unless noted otherwise, *epistemic reductionism* and *explanatory reductionism* are used interchangeably throughout the book.

2 I thank Arnaud Pocheville for articulating this point.

2 Multiple Realization and Reductionism

2.1 Introduction

One good departure point for discussing the relationship between different organizational or compositional levels in the biological world is the multiple realization thesis (MRT). Broadly understood, the MRT is a metaphysical thesis concerning the relationship between properties (or states, events, phenomena, etc.) located at different levels. Although levels can refer to many different things,[1] the notion of level here is narrowly understood in terms of organizational or compositional relationships, whereby entities at one level compose entities at the next higher level (Wimsatt 1974, 1976, 1994; Bechtel 2007). However, one might wonder why we need to consider a metaphysical thesis in a book defending an epistemic thesis. Just be patient, the answer will unveil itself when the time comes.

The MRT can be traced back to Hilary Putnam's seminal argument against "mind-brain identity theory" (Putnam 1967). The idea was further elaborated by Jerry Fodor (1974) and has continued to the present. The MRT was first introduced as an objection to the reduction of mental kinds (properties, states, events, etc.) to physical kinds in the philosophy of mind but later expanded into many other areas, such as the philosophy of psychology, the philosophy of biology, the philosophy of neuroscience, and so on.

Very recently, the long-standing debate surrounding multiple realization (MR) has been substantially invigorated. Though many detailed scientific examples have been introduced and extensively discussed (Aizawa 2007, 2009, 2013; Gillett 2007, 2010; Fang 2020a), agreement has not been achieved. On the one hand, the existence of MR is still in dispute, because many authors either believe that MR is metaphysically impossible[2] or is at most only epistemologically plausible (Clapp 2001). On the other hand, even if many authors grant the existence of MR, they disagree on how widespread cases of MR are. For example, although some authors take a skeptical stance on MR, they admit that it occurs in only a few cases (Shapiro 2000, 2004, 2008; Shapiro and Polger 2012;

DOI: 10.4324/9781003148029-2

Polger and Shapiro 2016). Others think there might be more cases of MR than opponents have envisioned (Sullivan 2008; Fang 2020b). Still others believe that MR not only exists but is also widespread.[3]

This chapter aims to show that the phenomenon of MR is widespread in the biological world. To do this, Polger and Shapiro's criterion for MR, namely, their Official Recipe, is discussed. Although Polger and Shapiro claim that their criterion makes MR less common than their proponents envision, I argue that their criterion renders MR less rare than they suggest. Hence, my disagreement with Polger and Shapiro only concerns their assessment of the scope of their criterion, rather than the criterion itself. The layout of this chapter is as follows: Section 2.2 introduces Polger and Shapiro's criterion for judging when the phenomenon of MR arises. Section 2.3 clarifies several areas of consensus surrounding MR. Then, on the basis of Polger and Shapiro's criterion and with the mentioned consensus areas at hand, a mechanistic multiple realization thesis (MMRT) is developed in Section 2.4. Two examples are also examined in this section to show that MR is prevalent in the biological world in terms of the MMRT. Finally, Section 2.5 considers the potential implications of the MMRT for the reductionism versus antireductionism debate in philosophy of biology.

2.2 Polger and Shapiro' Official Recipe

Intuitively, the MRT seems quite plausible, and many ordinary cases appear to support it. For example, as has been routinely argued on one side of the debate, although both cephalopods and human beings feel pain, the underlying physical processes involved are substantially different; that is, both are in the same mental state or have the same mental property but with different accompanying physical structures (thus with different physical properties).

On closer examination, however, the conclusion that such cases are ones of multiple realizability may be too hasty. To see this, consider a parallel case from Shapiro (2000). Suppose there are two corkscrews, one composed of steel and the other of aluminum. On the higher level, one might say that, because a corkscrew is typically used for removing corks no matter what microstructures are involved, these two corkscrews have the same *property of being able to remove corks*. But when moving to the level of microstructure, disagreement arises. The pro-MRT side would say that since they have different microstructures, they have different micro-properties. Hence, in this case, given that the two corkscrews have the same property of removing corks at the higher level, MR ensues. However, the anti-MRT side would reply that, admittedly, there are indeed different micro-properties in these cases, but these are *irrelevant* to realizing the higher-level property of removing corks. Hence, a case of MR is challenged. This negative outcome also holds for the case of cephalopods and human beings because, as the reasoning goes, although there are different micro-properties here, they are *irrelevant* with respect to realizing the mental property *pain*. In short, the anti-MRT side holds that

> [s]teel and aluminum are *not* different realizations of a waiter's cork-screws because, relative to the properties that make them suitable for removing corks, they are identical. The fact that one corkscrew is steel and the other aluminum is no more a reason to characterize them as different realizations than the fact that one might be yellow and the other red.
>
> (Shapiro 2000, 645; author's emphasis)

Shapiro adds that, although the fact that one corkscrew is steel and the other is aluminum is irrelevant with respect to removing corks, other properties such as mechanical principles—that is, the way corkscrews create friction on corks—are relevant. That is, Shapiro admits that if two cork-screws differ in mechanical principles, say one is a waiter's corkscrew and another is a winged corkscrew, then they realize the property of removing corks in different ways—thus, this is a case of MR (2000, 644; also see Polger and Shapiro 2016, 61–67). This brings us to the crux of the debate: On what grounds can we legitimately say that properties P_1 and P_2 (instantiated in two entities) are distinct realizations of property *Q*?[4] As the cork-screw case shows, this question cannot be straightforwardly answered. We need a *principled* way to determine when a case of MR is present. In this spirit, Shapiro offers a *causally relevant differences* criterion,[5] stating that

> [m]ultiple realizations count truly as *multiple* realizations when they differ in causally relevant properties—in properties that make a difference to how they contribute to the capacity under investigation.
>
> (Shapiro 2000, 644; author's emphasis)

That is, without making causally relevant differences to the realized kind's property, two property instances P_1 and P_2 cannot genuinely count as different realizers. Much clarification and refinement has occurred since this somewhat abstract criterion was first proposed (Shapiro 2008; Polger 2009; Shapiro and Polger 2012). Recently, the criterion has been refined by Polger and Shapiro (2016), using what they call the Official Recipe for MR. The Recipe says that for a case to be MR it must satisfy the following four conditions:

a. P_1 and P_2 are of the same kind in model or taxonomic system S_1.
b. P_1 and P_2 are of different kinds in model or taxonomic system S_2.
c. The factors that lead the P_1 and P_2 to be differently classified by S_2 must be among those that lead them to be commonly classified by S_1.
d. The relevant S_2-variation between P_1 and P_2 must be distinct from the S_1 intra-kind variation between P_1 and P_2. (Adapted from Polger and Shapiro 2016, 68)

Conditions (a) and (b) jointly capture the requirement that, for MR to arise, two different realizers must realize the same property (or kind). Note that *sameness* and *differentness* are defined by different sciences; that is, the realizing properties are classified as the same or not the same using a taxonomic system S_2 in science B, and the realized properties are classified as the same or not the same using a taxonomic system S_1 in science A (ibid., 26–32). Condition (c) captures the idea that MR requires P_1 and P_2 "to be not merely different, but to be 'relevantly different'—to be different in ways that are relevant to their performing the same function" (ibid., 68). The case of corkscrews illustrates this idea: the waiter's and winged corkscrews are different in ways that are relevant to how they remove corks, while the steel and aluminum corkscrews are not different in these ways (ibid., 68). Finally, condition (d) captures the idea that for MR to arise, the differences among realizers must be more than only individual variations (e.g., different colors in the corkscrew case); in other words, the differences must be big enough to classify two realizers as different kinds in terms of the taxonomic system S_2 in science B (ibid., 69).

Notice that, since Polger and Shapiro regard the realization relationship as different from both compositional and constitutional relationships, they use P_1 and P_2 to refer to both the realizing and realized properties. In particular, they suggest that for a property P to realize another property Q, P must have the function *constitutive* of Q's, that is, the Q-function (ibid., 23).[6] This view seems to assume the *flat* view of realization in contrast with the *dimensioned* view—I return to this point in the next section. On the other hand, however, they add immediately that their theory of MR does not depend on the correctness of their view of realization, since some other view of realization may better capture core characteristics of realization (ibid., 23). In one footnote (footnote 4 in Chapter 6), they add that their approach to MR does not make any assumptions about the right account of realization (ibid., 122). Given this, I suggest in the next section a compositional alternative to the conception of realization, which I believe better captures the kernel of the realization relationship.

With the Official Recipe at hand, we are now in a good position to evaluate when there is a case of MR and when there is no such case. However, before evaluating would-be cases of MR, we need to clarify several aspects concerning the concept of realization. This is the task of the next section.

2.3 The Conception of Realization

This section attempts to clarify the following aspects of realization: (a) realization is a compositional relationship, (b) realization involves different taxonomic systems defined by different sciences, (c) the abstract way of talking about realization calls for flesh and blood and (d) there is no interlevel causal relationship associated with realization.

2.3.1 A First Consensus: Realization Is a Compositional Relationship

Although a few authors have interpreted the realization relationship as a causal one (e.g., Sullivan 2008), disputants have, from the outset, described the relationship as noncausal, that is, as compositional or constitutional (Putnam 1967; Block and Fodor 1972; Fodor 1974; Boyd 1980; Kitcher 1984; Heil 1992, 1999; Bechtel and Mundale 1999; Sober 1999; Wilson 2001, etc.).

For current disputants, it is also clear that the realization relationship is compositional. For example, Kenneth Aizawa (2007, 2009, 2013) and Carl Gillett (2002, 2003, 2007) have argued extensively that realization is a *compositional/constitutive* relationship rather than a causal one.[7] Carl Craver (2004, 967–968) agrees, adding that compositional relationships have two forms: material/entity and property/activity realization, the latter of which can be further divided into aggregate[8] and mechanistic[9] realization. Craver and William Bechtel (2007, 550) also assert that a mechanistic relationship is a species of compositional relation. Craver (2004) says:

> In cases of *mechanistic realization*, the burden of realization is borne by some constituents (working parts or components) more than others, and organization among the components figures increasingly in our description of the realizer…. These components are organized such that they produce the behavior of the mechanism as a whole.
> (968; author's emphasis)

Given this consensus by both sides, we may say that the realization relationship should be better understood within a compositional framework. Yet, as mentioned in the last section, Polger and Shapiro disagree with this consensus. For them, realization differs from compositional and constitutional relationships. In particular, they claim that realization is a kind of *ontological dependence* relationship, meaning that "the dependent thing would not exist (or would not exist as the kind of thing it is) without the existence of that on which it depends" (Polger and Shapiro 2016, 19–20). For example, "a desk would not exist if the desktop and legs did not exist—the existence of the desk depends on the existence of its parts, the desktop and legs" (ibid., 20). Moreover, the realization relationship is *synchronic* (in contrast with the causal relationship which is diachronic) and *constant* (for instance, "current psychological states ontologically depend on current brain states and do so continually for as long as they persist" [ibid., 20]).

However, I prefer to see the realization relationship as compositional for the following reasons. To begin with, and most important, the reader may find that the key characteristics attributed to the realization relationship by Polger and Shapiro such as ontological dependence, synchronic and constant are also core characteristics we find in the compositional relationship. For example, it is legitimate to say that a desk is composed of its desktop and legs (such that the desk would not exist without the existence of the desktop and legs) and that this compositional relationship is clearly synchronic and constant. One should not be surprised by their sharing the same characteristics, because, as Aizawa and Gillett (2009a, 183) hold, the compositional relationship is just one type of realization relationship.[10] Second, since we restrict our focus to the biological world, which is typically characterized as a hierarchy of organizational levels (e.g., an organism is composed of its tissues, which are composed of organelles, which, in turn, are composed of cells, etc.; see Wimsatt 1974, 1976, 1994; Bechtel 2007), I think it is appropriate to view the realization relationship in the biological world through the compositional lens.

Given the choice (i.e., seeing the realization relationship as compositional), the work to be done can be narrowed down to an investigation of the scope of multiple compositional realization in the biological world.

2.3.2 A Second Consensus: Realization Involves Different Taxonomic Systems

As has been made very clear by Polger and Shapiro, evaluating cases of MR involves comparing different taxonomic systems defined by different sciences. That is, as discussed in Section 2.2, to count as a genuine case of MR, the realizing properties must be classified as different kinds by a taxonomic system S_2 in science *B*, and the realized properties must be classified within the same kind by a taxonomic system S_1 in science *A* (Polger and Shapiro 2016, 26–32). Polger (2008) claimed that "[m]ultiple realization occurs when items of the same special science kind fall into different base science kinds" (539) and that "to find out which differences are relevant we must attend to the *sciences* that are *in the business of classifying S's and B's*, respectively"[11] (545; my emphasis). Mark Couch (2004), another objector to the MRT, agrees that "[e]stablishing multiple realization requires showing that the same *function* is realized by different types of physical states" (202; my emphasis). Note that these disputants seem to agree that the realized properties can be broadly construed as functions or functional kinds, where a function denotes what a property, state, kind, or object can do (Polger and Shapiro 2016, 22).

Interestingly, this construal suits the pro-MRT side's taste. Charitably read, Fodor's central idea concerning MR is not so much that the list of realizing properties is so open-ended that a higher-level property can be multiply realized but that lower-level properties *defined by lower-level sciences* are of disparate kinds while higher-level properties *defined by higher-level sciences* are of supposedly homogenous kinds (Richardson 2008, 530). Robert Richardson (2008), another proponent of the MRT, says that

> [i]n the end, lower-level properties are not the same *kinds*, construed in terms of theories pitched toward the lower level. However, assuming there is an articulated and defensible higher-level theory, the corresponding higher-level properties are of a unified—perhaps homogeneous—kind, in terms articulated by the appropriate higher-level theory.... Much depends here on talk of levels of organization and on talk of levels of theory.
>
> (530; author's emphasis)

In sum, the consensus between both sides boils down to the simple claim that there is a case of MR when and only when a (functionally defined) property/kind *Q* classified by a taxonomic system in a higher-level science/theory *Y* can be multiply realized by distinct properties/kinds P_1, P_2, ..., P_n classified by a taxonomic system in a lower-level science/theory *X*. For simplicity, in what follows, I only claim that for there to be a case of MR, a property *Q* in a higher-level science *Y* can be multiply realized by distinct properties P_1, P_2, ..., P_n in a lower-level science *X*. However, as we shall see in the next subsection, this way of putting the consensus is still too abstract, for it is not quite precise to say that a property *P* realizes a property *Q*.

2.3.3 The Flesh-and-Blood Form of Mechanistic Realization

The third point crucial for understanding realization concerns the *flesh-and-blood* form of MR. It is characteristically claimed that property *P* realizes property *Q* (Putnam 1967; Fodor 1974; Kim 1992; Shapiro 2000, etc.). In light of the mechanistic relationship discussed above, we now know that this is an imprecise way of talking about realization, for *Q* is either an abstraction from its organized components (including components, their properties and their organization) or the sum of the properties of the components. If the former, then *P* and *Q* are in fact identical, for the property of being a certain arrangement of organized components is just the property of the whole, that is, *Q*. If the latter, then saying that *Q* is the sum of the properties of the components is simply wrong, because there is no such thing as the sum of the properties of the components in the case of mechanistic realization (although this might be true in the case of aggregate realization).[12] For instance, a diamond's property of being hard is not the sum of the properties of its components, namely carbon atoms. Given this, I think it is fruitful to propose an alternative approach to discussing realization: we will say that *an instance*, *entity*, or *individual S*—rather than a property—composed of entities e_1, e_2, ..., e_n, which are organized in a certain way *M*, realizes the property *Q*. e_1, e_2, ..., e_n could be the same or different sub-entities and have their own properties p_1, p_2, ..., p_n, respectively. Note that the relata of realization is no longer between two properties but between an individual and a property; it is a certain individual that realizes a certain property.[13] The idea is expressed in Figure 2.1.

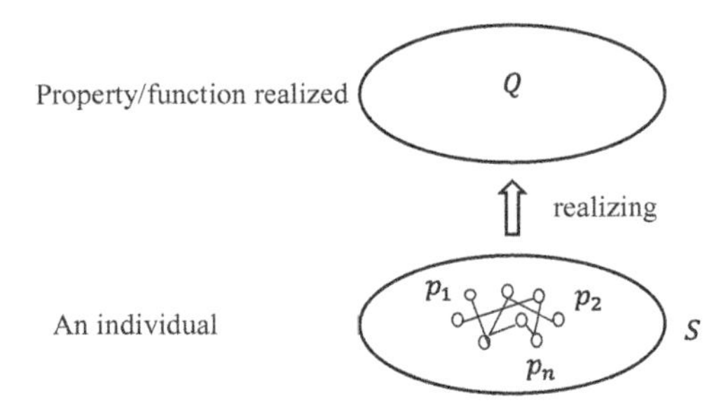

Figure 2.1 An individual S made up of components $e_1, e_2, \ldots, e_n$ realizes the property/function Q.

Figure 2.1 uses two big ellipses to describe component properties and realized properties p_i and Q, respectively, to highlight the idea that p_i and Q are not necessarily instantiated in the same individual S. For example, p_1 may be instantiated in sub-entity e_1 but not in individual S, while Q may be only instantiated in S but not in e_1. Furthermore, the lower ellipse in Figure 2.1 contains a number of small ellipses standing for the components of the individual, indicating that these elements may have the distinctive properties $p_1, p_2, \ldots, p_n$ (and the distinctive causal powers) that cannot be found in individual S.

This alternative approach to discussing realization reminds us of the *dimensioned view* suggested by Carl Gillett (2002, 2003). The dimensioned view is contrasted with the *flat view*, which is also called the standard view because it is accepted as a default position among major disputants (e.g., Kim 1998; Shoemaker 2001). The flat view has two central tenets: (a) "a property instance X realizes a property instance Y only if X and Y are instantiated in the same individual" and (b) "*only if* the causal powers individuative of the instance of Y match causal powers contributed by the instance of X (and where X may contribute powers not individuative of Y)" (Gillett 2002, 317–318; author's emphasis). In short, the flat view takes "realized and realizer properties to share both the individual in which they are instantiated and at least some of the causal powers contributed to this individual" (ibid., 318).

In contrast, the dimensioned view runs contrary to both of these tenets. First, the dimensioned view holds that the realized and realizer properties are not necessarily instantiated in the same individual. For example, a diamond has the property of hardness, while components of the diamond, that is, carbon atoms, have the properties/relations of being bonded and aligned with one another.[14] Hence, the property of hardness is only instantiated in the diamond whereas the properties/relations of being bonded and aligned are only instantiated in the carbon atoms. Second, the dimensioned view holds that the causal powers of the realized property do not necessarily match the causal powers of the realizer properties. For instance, the diamond has the causal power of scratching glass, while the carbon atoms have the power of causing a contiguous carbon atom to remain in a certain position and in a certain direction. Therefore, the causal power of the diamond does not match the causal powers of the carbon atoms.

However, although one may find that the dimensioned view has a clear bearing on the alternative approach to discussing realization suggested in this section, differences between them remain. As shown earlier, the dimensioned view sees the properties instantiated in components of an individual as realizer properties (e.g., properties of carbon atoms). In contrast, the alternative approach views the individual itself as the realizer, and the components of the individual are not themselves realizers—although the way these components are organized is relevant to realizing the realized property. With this clarification, the dispute about whether the realizer and realized properties are instantiated in the same individual and whether the causal power of the realized properties matches the causal power of the realizer properties disappears. Now, it is the individual, as *the* realizer, that realizes a certain property, not a property instantiated in one individual (typically a component) that realizes another property instantiated in the same or a different individual.[15] Therefore, the alternative approach differs from both the flat and the dimensioned views of realization—although it bears a closer affinity to the dimensioned than to the flat view (for it draws on the dimensioned view in taking realization to have something to do with how a system is made up of its components).

Note that, although it is correct to say that an individual realizes a property, we must look into the details of the individual in order to understand how a property is realized. An individual is made up of components organized in a certain way. That is, the realizer consists of components, properties of these components and a particular organization thereof—for this reason, sometimes I simply use the term *organized components* to denote the individual. As such, the problem of how an individual realizes a property consists in how its components are organized in a certain way so as to realize a certain property. As a consequence, MR consists in how a property in question can be realized differently, for example, realized by two different organizations of components. If the organized components turn out to realize the property under consideration in different ways (e.g., different organizations that matter to producing the overall function of the individual), then there is a case of MR. So, to judge whether a property is multiply realized by two (or more) individuals, we first must decide if the components involved in these two individuals are classified as different or not and if the organizations of the components of the two individuals are classified as the same or not. We must then decide if the differences in the realized properties are only intra-kind variations or inter-kind disparities. Finally, we must make sure of whether the differences in components and/or the differences in organizations matter to how the realized properties are realized, that is, if they are relevant differences.

2.3.4 The Interlevel Relationship

The final crucial point necessary for understanding realization is that many authors following the compositional approach are fully aware that there is no interlevel causal relationship (Aizawa 2007, 2009, 2013; Gillett 2002, 2003, 2007). For example, Craver and Bechtel (2007) contend explicitly that "talk of interlevel causation is merely a misleading way to talk about an explanatory interlevel relationship that, upon close inspection, does not involve interlevel causes" (562). They admit, however, that causal relationships do exist within intralevel components. That is, components organized in certain ways causally interact with one another and thus constitute the property/activity of the whole.

Given these considerations—especially the first and second consensuses and the *flesh-and-blood* form of mechanistic realization—the next section considers how we can integrate Polger and Shapiro's criterion with these considerations to build a more detailed account of MR and how to use this more detailed account to evaluate would-be cases of MR in the biological world.

2.4 Multiple Mechanistic Realization and Case Studies

Recall that the core of Polger and Shapiro's Official Recipe is that, for there to be a case of MR, two realizers must be classified as different kinds by a taxonomic system S_2 in science B, and the realized properties must be classified within the same kind by a taxonomic system S_1 in science A. Also recall the first consensus that realization is a compositional relationship, the second consensus that realization involves different taxonomic systems and the *flesh-and-blood* form of MR (i.e., that an individual S composed of entities $e_1, e_2, ..., e_n$, organized in a certain way M, realizes the property Q). Combining all these elements together, we obtain a detailed account of MR, which I call the multiple mechanistic realization thesis (MMRT):[16]

(MMRT): a case is an example of multiple mechanistic realization, *iff*:

1. there are at least two individuals S_1 and S_2 that are classified into different kinds by a lower-level science X;[17]
2. S_1 is composed of entities $e_1, e_2, ..., e_n$, organized in a certain way M, S_2 is composed of entities $f_1, f_2, ..., f_n$, organized in a certain way N and $e_1, e_2, ..., e_n \neq f_1, f_2, ..., f_n$ (meaning that the set $\{e_1, e_2, ..., e_n\}$ is distinct from the set $\{f_1, f_2, ..., f_n\}$), and/or $M \neq N$;
3. S_1 and S_2 manifest the same property Q or *similar* properties Q_1 and Q_2, respectively;
4. there is a higher-level science Y ($Y \neq X$) that classifies Q_1 and Q_2 as being of the same kind Q and classifies S_1 and S_2 as members of the same kind S in virtue of property Q (or similar properties Q_1 and Q_2) they manifest; and

5. the differences between $e_1, e_2, \ldots, e_n$ and $f_1, f_2, \ldots, f_n$, and/or between M and N are *relevant* to how S_1 and S_2 realize Q (or Q_1 and Q_2).

Obviously, the MMRT satisfies all conditions of Polger and Shapiro's criterion: (I) S_1 and S_2 are of the same kind in a higher-level science Y, (II) S_1 and S_2 are of different kinds in a lower-level science X, (III) the factors that lead the S_1 and S_2 to be differently classified by X are those that lead them to be commonly classified by Y, because it is the *relevant differences* between $e_1, e_2, \ldots, e_n$ and $f_1, f_2, \ldots, f_n$, and/or between M and N that lead the S_1 and S_2 to be differently classified by X (since these differences lead to inter-kind disparities in X) and to be commonly classified by Y (since these differences merely lead to intra-kind variations in Y); and (IV) as implied by (III), the relevant X-variations between S_1 and S_2 are inter-kind disparities, whereas the Y-variations between S_1 and S_2 are intra-kind variations, so these are two distinct types of variations.

In addition, the MMRT integrates all the considerations articulated in the last section. For example, it treats realization within the compositional framework (e.g., S_1 is composed of $e_1, e_2, \ldots, e_n$ and these components and their organization matter to how an individual realizes a property), takes into account the fact that MR involves different taxonomic systems (e.g., X and Y), incorporates the flesh-and-blood consideration (i.e., it is an individual made up of components that realizes a property) and implies that there is no inter-level causation (i.e., it is the organized components that *realize* rather than *cause* a property).

Looking closely, it may be that the MMRT implies three different ways of obtaining multiple mechanistic realization: (MMR1) if S_1 and S_2 are composed of two different kinds of entities and these different kinds of entities are organized in the same way, and the different kinds of entities matter to how S_1 and S_2 *differently* realize a *common* property Q; (MMR2) if S_1 and S_2 are composed of the same kind of entities and these entities are organized in different ways, and the different organizations matter to how S_1 and S_2 *differently* realize a *common* property Q; and (MMR3) if S_1 and S_2 are composed of different kinds of entities and these different kinds of entities are organized in different ways, and both the different kinds of entities and different organizations matter to how S_1 and S_2 *differently* realize a *common* property Q.

Piccinini and Maley (2014, 137–141) also argue that we have these three types of MR and that all three count as genuine cases of MR insofar as the differences in the realizer (i.e., components and/or organizations) are relevant to the realization of the realized property. I agree with them and think that examining the second type (i.e., MMR2) is sufficient for our current purposes, namely, establishing that MMR is not difficult to find in the biological world.[18] Hence, I only consider would-be cases of MMR2 in what follows. The way to do this is to see whether MMR2 can be readily found in the biological world as investigated by various biological sciences, depending on different levels and aspects of analysis. Two cases are examined, one concerning neural plasticity in neuroscience and another concerning isozymes (also known as isoenzymes) in biochemistry.

2.4.1 Neural Plasticity

This example comes from Polger and Shapiro (2016, 90–98). The reason for reexamining this example is simply that I think Polger and Shapiro's analysis of the example is problematic. Let us turn to their example. There is one form of neural plasticity called cortical functional plasticity, referring to the phenomenon in which "whole areas of the cortex seem to perform different tasks at different times or in different subjects" (ibid., 90). One case in point is that of "rewired" ferrets:

> The standard pathway from the ferret's eye to its visual cortex travels through the lateral geniculate nucleus and the lateral posterior nucleus. Von Melchner et al. (2000) redirected retinal axons from the right visual field that usually project to these areas, connecting them instead to the medial geniculate nucleus, which innervates the audio cortex. Thus, the 'rewired' ferret's auditory cortex received visual information from the ferret's right visual field.
>
> (Polger and Shapiro 2016, 92)

One important difference between the visual and auditory cortexes is that they show substantially different forms of organization:

> Within visual cortex, groups of cells are arranged into orientation columns, with each column of cells especially tuned to specific orientations of stimulus (Sharma et al. 2000). Moreover, the visual cortex contains a 2D map of the retina, with each point on the retina corresponding to a point on this map. In contrast, the auditory cortex contains no columns of orientation-sensitive cells. Nor does it contain a 2D map of auditory space. Rather, the cochlea maps onto a 1D map in auditory cortex and neurons in auditory cortex are grouped in clusters that receive excitation from both ears, or excitation from one ear and inhibition from the other (Roe et al. 1990).
>
> (Polger and Shapiro 2016, 92)

One surprising thing happened after scientists performed the operation on the visual and auditory cortexes: the "rewired" ferrets regained vision in their right visual fields:

> The ferrets that had been trained to respond in one way to an auditory stimulus and another to a visual stimulus displayed the visual

> response to stimuli presented to their right visual fields (von Melchner et al. 2000). Rewired ferrets were also tested for their visual acuity, and they were able to discern gratings of various frequencies and at various contrasts. In short, the rewired ferrets appear able to see with their auditory cortex.
>
> (Polger and Shapiro 2016, 92–93)

So far, the case being discussed suggests that this is a case of MMR, for it satisfies all conditions of MMR2. That is, (1) the visual and auditory cortexes are classified as being of two different kinds in a lower-level science (e.g., neuroscience); (2) the visual and auditory cortexes have substantially different organizations; (3) the visual and auditory cortexes manifest the same or similar visual capacities; (4) going one level higher, the visual and auditory cortexes are classified as being of the same kind (suppose, e.g., there is such an "eyelike"-kind defined by visual capacities) in terms of their common visual capacities; and (5) the differences between the organization of the two cortexes are relevant to how they realize the respective visual capacities. Assessing whether our case meets condition (5), however, requires further analysis. The rewired ferrets were able to see with their auditory cortex but did not do so as perfectly as the normal ferrets. For example, when placed under test conditions, "the rewired ferrets show significant degradation in their discriminatory ability, being unable to detect gratings at lower contrasts or higher spatial frequency than the normal ferrets" (Polger and Shapiro 2016, 95). This difference in discriminatory ability is due to the fact that the auditory cortex of the rewired ferrets is an imperfect machinery with respect to *seeing things*, and it is the internal organization of the auditory cortex that makes machinery less perfect than the visual cortex of the normal ferrets.

However, Polger and Shapiro do not buy my verdict. Their argument has two main parts. To begin with, due to the difference in discriminatory ability of the rewired and normal ferrets, Polger and Shapiro think it is illegitimate to classify their visual capacities as being of the same kind. Rather, the visual capacities of the rewired and normal ferrets belong to two different kinds, so that the case in question is not a case of MR. They claim that "[i]f visual processing in the visual and auditory cortexes were indeed the same, we should expect the normal and rewired ferrets to perform *identically* in discrimination tasks" (ibid., 95; my emphasis). I am inclined to disagree. First, recall Polger and Shapiro's Official Recipe, according to which the realized property or kind should tolerate intra-kind variations. A consequence is that it is not to be expected that all members of a kind manifest exactly the same or *identical* property (and I doubt that all members of a kind could ever manifest exactly the same or identical property). The intra-kind variation is a matter of degree, while being identical or not is an all-or-nothing issue. Therefore, insofar as the rewired ferrets had visual capacities, although not as perfect as the normal ferrets, it remains unclear why we should not classify them as being of the same kind as the normal ferrets. Second, aside from the type of intra-kind variations in the rewired ferret case, other types of intra-kind variations are numerous. For example, a very old ferret's visual capacities might not function as perfectly as that of a younger ferret (e.g., the old ferret, when performing the test, is unable to detect gratings at a lower contrast or higher spatial frequency). Also, a shortsighted ferret's visual capacities may well be remarkably worse than that of a normal-sighted ferret, with the shortsighted ferret performing very poorly in detecting gratings at lower contrasts or higher spatial frequencies than the normal-sighted ferret and so on. Hence, if we classify the visual capacities of the rewired and normal ferrets as being of different kinds, nothing can prevent us from classifying the visual capacities of the old and younger ferrets as being of different kinds and classifying the visual capacities of the shortsighted and normal-sighted ferrets as being of different kinds.

Polger and Shapiro's second major argument is that, since there is a tendency that the "the auditory cortex in the rewired ferrets appears to be trying its best to turn itself into a visual cortex for purposes of processing the information it receives from the retina" (ibid., 97–98), the differences in the rewired and normal ferrets may be not large enough to classify them as being of different kinds. In particular,

> [w]hat one notices when examining the auditory cortex of a rewired ferret is a cortex structured something like a normal auditory cortex, but also something like a normal visual cortex (Sharma et al. 2000). Thus, the rewired auditory cortex displays columns of orientation-sensitive cells, just as normal visual cortex does. It also contains regions that bear horizontal connections to each other, as do regions in visual cortex.
>
> (Polger and Shapiro 2016, 96)

Nonetheless, the rewired auditory cortex is not exactly the same as the normal visual cortex, for

> whereas one finds a great number of orientation maps in visual cortex that resemble something like a pinwheel, in rewired auditory cortex the density of these maps is far lower. Moreover, the volume of the regions of orientation-sensitive cells in rewired auditory cortex is much larger than the volume of these regions in visual cortex.
>
> (ibid., 96)

Given these similarities and dissimilarities, Polger and Shapiro claim that

> [t]o the extent that they do realize visual processing, they also resemble the structure of visual cortex. Correlatively, to the extent that

their physical organization diverges from visual cortex, they fail to process visual information as successfully as visual cortex does.
(ibid., 98)

In other words, the similar visual capacities of the rewired and normal ferrets occur due to their similar organizations, and the variations of their visual capacities are due to their organizational variations. Therefore, this is not a case of MR, since the criterion of one property or function realized in two different ways is not met.

However, the devil is in the details. To see this, let me examine Polger and Shapiro's example more closely. Polger and Shapiro claim that there are organizational similarities between the rewired and normal ferrets, because the rewired auditory cortex also displays columns of orientation-sensitive cells and contains regions that bear horizontal connections to each other. There are two features that influence the organization of the cortex: orientation preference and horizontal connections. With respect to the first feature, Sharma *et al.* (2000, 842) state that the two cortexes contain a pinwheel organization, although the density of pinwheel centers in the rewired auditory cortex is one-quarter of that present in the visual cortex. With respect to the second feature, however, the situation becomes more complex. It is true that the rewired auditory cortex also contains regions that bear horizontal connections to each other, but key features of horizontal connections such as cell aggregation and sizes of cell patches differ in these regions in the rewired auditory and the normal visual cortexes. Cell aggregation denotes the extent to which cells are clumped into patches or are randomly distributed (Sharma *et al.* 2000, 845). Sharma *et al.* show that cell aggregation in the rewired auditory cortex is larger than in the normal auditory cortex but smaller than in the normal visual cortex (ibid., 845). That is, the rewired auditory cortex stands somewhere in between the normal auditory and visual cortexes with respect to the property of cell aggregation. In addition, the size of cell patches in the rewired auditory cortex is larger on average than that of the normal visual cortex but smaller than that of the normal auditory cortex (ibid., 845). In other words, this feature of the rewired auditory cortex also differs from that of both the normal visual and auditory cortexes.

In short, the rewired auditory cortex differs from both the normal visual cortex and normal auditory cortex due to the differences in the two sub-features of horizontal connections: cell aggregation and the sizes of cell patches. Given this, it seems that the rewired auditory cortex stands somewhere in the middle of the normal visual and normal auditory cortexes. This conclusion is in concert with Polger and Shapiro's (2016) claim that "[w]hat one notices when examining the auditory cortex of a rewired ferret is a cortex structured something like a normal auditory cortex, but also something like a normal visual cortex" (96). But if the rewired auditory cortex stands in the middle of the two cortexes, then there is no reason, without further inquiry, to classify it as being of one kind (i.e., the kind including the normal visual cortex) rather than the other (i.e., the kind including the normal auditory cortex).

The argument presented above can be reinforced if we take a closer look at the sub-features of horizontal connections. Among the two key sub-features, cell aggregation is of great importance. As stated earlier, cell aggregation refers to the extent to which cells are clumped into patches or are randomly distributed. Sharma *et al.* (2000) use the *index of cell aggregation* to measure cell aggregation, which requires quantifying clustering in cell plots, that is, measuring the degree to which cells are clustered together.[19] Since neural cells (or neurons) are connected with one another (directly or indirectly) and it is a basic practice in science to represent linked neurons as neural networks,[20] an alternative way to measure cell aggregation is to use tools from network biology.

In network biology, the clustering coefficient $C_I = 2n_I/k_I(k_I - 1)$ is an important topological characteristic[21] used to measure the degree to which cells in a network are linked together (in particular, linked as triangles). More precisely, in the coefficient, given a cell I, n_I refers to the actual number of connected pairs of neighbors of I and $k_I(k_I - 1)/2$ to the maximum possible number of connected pairs of neighbors of I, should all of cell I's neighbors be connected to each other (Barabási and Oltvai 2004, 102; also see Kantarcı and Labatut 2013).[22] Figure 2.2 provides an example of a simple network.

This network has eight nodes (representing cells) and 10 edges (representing relations between cells). In this network, only the pair B and C of node A's five neighbors are connected together, so $n_A = 1$ and the clustering coefficient $C_A = 2/20 = 0.1$ (Barabási and Oltvai 2004, 102). So far, we have only described the clustering feature of a single node. To describe the global tendency of nodes in a network to form clusters, we require an average clustering coefficient, namely, the $<C>$ (ibid., 102), which is

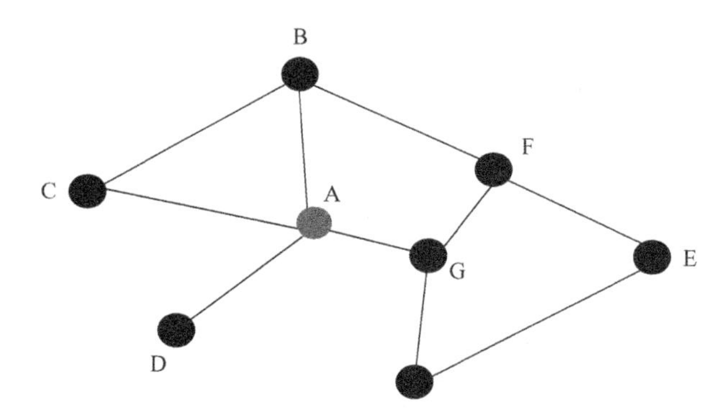

Figure 2.2 Undirected network. This figure comes from Barabási and Oltvai (2004, 102) with minor modifications. Figure used with permission.

simply the ratio of the sum of the clustering coefficient of all nodes to the number of nodes (Watts and Strogatz 1998). Based on the <C>, we obtain another important measure of a network's structure (also called topology),[23] namely, the function $C(k)$, which is "defined as the average clustering coefficient of all nodes with k links" (ibid., 102).[24]

We can now use the tools described earlier to measure cell aggregation. Since cell aggregation in the rewired auditory cortex is smaller than in the normal visual cortex and larger than in the normal auditory cortex, it can be expected that when using the $C(k)$ to measure the structural features of the three cortexes, three different images will emerge; that is, the rewired auditory cortex will show a topology that differs from both that of the normal visual and that of normal auditory cortexes.[25] In other words, the rewired auditory cortex possesses a distinct topological characteristic, and this enables us to reason that it constitutes a sort of network that differs from both the normal visual and normal auditory networks. Moreover, this difference in topological characteristics matters for functioning, connecting or collaborating with other networks and even the evolution of the networks in question (Barabási and Oltvai 2004). For this reason, I think it is illegitimate to classify the rewired auditory and normal visual cortexes as being of the same kind from either the perspective of network biology or the perspective of neuroscience employing the machinery from network biology. Instead, a better strategy is to classify the rewired auditory cortex as being of a distinct kind or, to put it in a different way, classify it as being of an intermediate kind that differs from both these two kinds.

In sum, we have seen that Polger and Shapiro's two objections to the example of rewired ferrets classifying as a case of MR fail. For one, since the variations of visual ability in the rewired auditory and normal visual cortexes are merely intra-kind variations in a science of "eyes", it is better to consider the two cortexes as being of the same kind. For two, because the rewired auditory and normal visual cortexes constitute two distinct neural networks in terms of their respective topological characteristics, it is better to classify them as being of different kinds through the lens of neuroscience. Therefore, the example is a case of MR (and MMR2 in particular).

2.4.2 *Isozymes*

We can now consider a case concerning isoenzymes from biochemistry, one that also instantiates MMR2. Let us start with a definition of isoenzymes. A textbook definition states that

> [*i*]sozymes or *isoenzymes*, are enzymes that differ in amino acid sequence yet catalyze the same reaction. Usually, these enzymes display different kinetic parameters, such as K_M, or different regulatory

> properties. They are encoded by different genetic loci, which usually arise through gene duplication and divergence.
>
> (Berg *et al.* 2002, Section 10.3; author's emphasis)

Gunning (2001) states that

> [i]soforms are highly related gene products that *perform essentially the same biological function*. Isozymes are isoforms of an enzyme. Isoforms can differ in their biological activity, regulatory properties, temporal and spatial expression, intracellular location or any combination thereof.
>
> (1; my emphasis)

In other words, isozymes are enzymes (proteins)[26] with different structures but the same biological function—that is, they are *functionally equivalent*. Functional equivalence is common in isoforms: "the function of isoforms goes beyond differences in their polypeptide chains.... Evolution selected multiple engrailed genes to provide necessary gene regulatory programs and has tolerated protein divergence within the bounds of apparent functional equivalence" (Gunning 2001, 3). Enzymes are functionally defined entities and can be classified into six groups (and a greater number of subgroups) based on the *types* of reaction they catalyze: *oxidoreductases*, *transferases*, *hydrolases*, *lyases*, *isomerases*, *ligases*.

Consider the following example. Glucokinase and hexokinase I are typical examples of isozymes. Hexokinase has four isozymes: hexokinases I, II, III, and IV, which are homologous[27] (Iynedjian 1993, 6). Animals deploy a wide range of hexokinases. In particular, hexokinase I predominates in brains, whereas hexokinase IV, also known as glucokinase, is found mainly in the liver and the pancreas (Smith 2000, 171). Hexokinase I and glucokinase are viewed as isozymes because they both catalyze the phosphorylation of glucose (Figure 2.3).

Although they catalyze the same reaction (or realize the same function), their structures are not exactly the same:

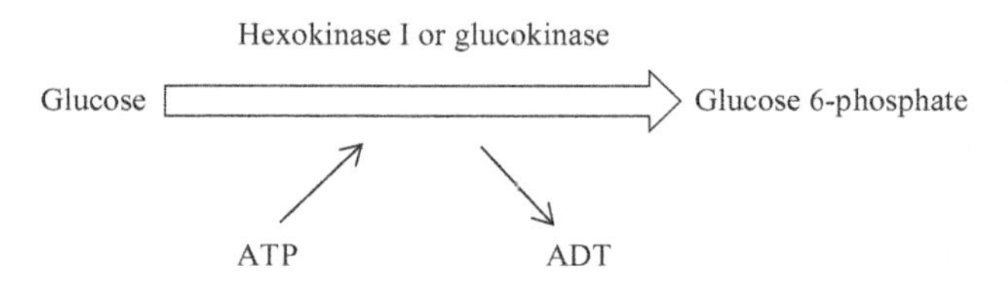

Figure 2.3 Metabolism of glucose.

> The Types I, II, III isozymes, and the tumor enzyme, consist of a single polypeptide chain with a molecular mass of ~100 kDa, and all four are inhibited by the product Glc-6-P (6). The Type IV hexokinase (also known as glucokinase), similar to yeast hexokinase, is ~50 kDa and is insensitive to inhibition by Glc-6-P (6).
>
> (Arora *et al.* 1993, 18259)

The structural difference is partly due to the fact that glucokinase is coded by a different gene (Kanno 2000, 83). It should come as no surprise that the structural difference, in turn, causes differences in biological activity, regulatory properties, temporal and spatial expression, intracellular location, and the like. For instance, glucokinase has a lower affinity for glucose than the other hexokinases; the lower affinity for glucose means that glucokinase functions only when serum glucose levels (i.e., concentration) are high (Iynedjian 1993, 7). In addition, glucokinase is not inhibited by its product (viz. glucose 6-phosphate), which allows it to remain active in storing as much glucose as possible in the presence of high glucose levels. By contrast, other hexokinases are inhibited by glucose 6-phosphate when it reaches a certain level (ibid., 7). However, these differences are better conceptualized as *intra-kind variations*, because hexokinases are defined as functionally equivalent with respect to the property of catalyzing the phosphorylation of glucose.

It can be argued that this example also counts as a case of MMR, for it meets all conditions of MMR2. More specifically, (1) hexokinase I and glucokinase are classified as being of two different kinds in a lower-level taxonomic system due to their differences (or inter-kind disparities) in biological activity, regulatory properties, and so on; (2) hexokinase I and glucokinase have different organizations (i.e., structures); (3) hexokinase I and glucokinase manifest the same functionally defined property of catalyzing the phosphorylation of glucose; (4) going a level higher, hexokinase I and glucokinase are classified as being of the same kind in terms of their common property of catalyzing the phosphorylation of glucose (although there are intra-kind variations with respect to this property); and (5) the differences in the organizations of hexokinase I and glucokinase are relevant to how they realize the property of catalyzing the phosphorylation of glucose, because the organizational differences result in the *intra-kind variations* of the property; for example, glucokinase has a lower affinity for glucose and is not inhibited by its product.

Note that there is a difference between this example and the example of the rewired ferrets discussed in the last subsection. In the example of the rewired ferrets, we assumed that the rewired auditory and normal visual cortexes were made up of the same components (i.e., the same neural cells), although their respective compositions differed. In the example discussed here, there are in fact differences in the components due to differences in amino acid sequences (considered aside from differences in organization). That is, hexokinase I and glucokinase are composed of entities that are nonidentical. These differences in components may be relevant to how hexokinase I and glucokinase realize the property in question. However, for reasons of simplicity, and given that the differences in the organizations are sufficient to establish MMR, I leave to another occasion the question of whether the differences in the components of hexokinase I and glucokinase are relevant to how they realize the property.

To conclude, we have here another case of MMR associated with isozymes hexokinase I and glucokinase. Furthermore, given that isozymes constitute a large group of substances in the world, it should not be surprising to generalize that we have found a large kind of cases of MMR, rather than a single case of MMR. Likewise, we would say that the phenomenon of neural plasticity (perhaps, more broadly, biological plasticity) constitutes another kind of MMR.[28] Though the list of the kinds of phenomena that could possibly constitute cases of MMR can be further extended, the two kinds discussed so far are sufficient to indicate that there is no reason to claim that MR is not as common as philosophers usually envision.

2.5 Implications for the Reductionism Versus Antireductionism Debate

So far, I have shown that MMR is not so difficult to find in the biological world. Since the goal of this book is to prove the explanatory autonomy of the biological sciences, and the MR dispute is usually linked to the reductionism versus antireductionism debate in the philosophy of biology (e.g., Sober 1999; Rosenberg 2001a, 2001b, 2001c, 2006), in the following section and chapter, I consider the implications of the MMRT for the reductionism debate in the philosophy of biology.

There are different conceptions of reductionism. For example, some authors distinguish ontological reductionism from epistemic reductionism and theory reductionism (e.g., Sarkar 1992). Some distinguish ontological from methodological and epistemic reductionism, with theory and explanatory reductionism considered as two forms of epistemic reductionism (Brigandt and Love 2015). Still others distinguish ontological from epistemic reductionism, with explanatory and methodological reductionism seen as subtypes of epistemic reductionism (Hüttemann and Love 2011, 524). Putting these differences aside, theorists agree that there is, at the very least, a meaningful distinction to be drawn between ontological and epistemic reductionism. Throughout this book, I assume this distinction. In addition, theorists appear to agree that explanatory reductionism is a subtype of epistemic reductionism, hence this book will also assume the existence of this subcategory.

Metaphysical (or ontological) reductionism refers to the stance that "each particular biological system (e.g., each individual organism) is constituted by nothing but molecules and their interactions" (Brigandt and Love 2015). Epistemic reductionism denotes the position that "the knowledge about one scientific domain (typically about higher level processes) can be reduced to another body of scientific knowledge (typically concerning a lower and more fundamental level)" (ibid.). Explanatory reductionism holds that any higher-level feature can be explained by representations (e.g., generalizations, explanations, models, etc.) of lower-level features (ibid.) and that the explanatory significance is primarily, if not completely, located at the lower level (Wimsatt 1974; Waters 1990, 1994, 2000, 2008; Sarkar 1992, 1998, 2001, 2002, 2005; Weber 2005). The next chapter discusses the debate surrounding explanatory reductionism at length.

Given these distinctions, let us now return to our question: What are the implications of the MMRT for the reductionism debate? We now know that the answer to this question depends on what we mean by the term reductionism. If we mean metaphysical reductionism, then the MMRT seems to be compatible with the stance—although some philosophers interpret MR as a thesis against the *type-type* version of metaphysical reductionism, for example, Putnam (1967) and Block and Fodor (1972), among others.[29] The MMRT is compatible with metaphysical reductionism because what the MMRT denies is that there is only one single way to constitute (or realize) a particular biological system—namely, it denies the unique realization thesis—rather than denying that each biological system is constituted by molecules and their interactions. In other words, the thesis that biological systems can be constituted by molecules and their interactions *in different ways* is compatible with the stance that each biological system is constituted by molecules and their interactions.

On the other hand, if what we have in mind is epistemic reductionism, then the question concerning the implications of the MMRT for the reductionism debate becomes more complicated. This is because, first, the standard way of advancing the MR debate is to conceive of it as a metaphysical thesis; that is, to conceptualize the thesis as a claim about the relationship between different properties (or states, events, phenomena, etc.) rather than the relationship between different bodies of knowledge or between different epistemic entities such as generalizations, theories, models, and so on. Thus, there seems to be a mismatch between the MMRT and the concept it is supposed to relate to, that is, epistemic reductionism.

Second, it seems that the epistemic reductionist position may not follow directly from metaphysical reductionism; that is, it is not the case that if metaphysical reductionism holds then epistemic reductionism holds (Brigandt and Love 2015). As said earlier, the MMRT is compatible with metaphysical reductionism. The remaining problem, then, is about the relationship between the MMRT and epistemic reductionism. In particular, since this book only concerns the explanatory autonomy of the biological sciences, the remaining problem concerns the relationship between the MMRT and explanatory reductionism, a subtype of epistemic reductionism.

Considered closely, we may find that the MMRT is neutral with respect to the explanatory reductionist position, for the MMRT says nothing about which form of explanation is better or where a better explanation resides, nor does it imply either—although some authors interpret the MR thesis as an argument against epistemic reductionism, for example, Hull (1972, 1974), Fodor (1974), Wimsatt (1976) and others. Instead, the MMRT only states that different systems can realize the same property or function. One may argue, however, that even although the MMRT does not say anything explicitly about scientific explanation, it may have implications for the explanatory reductionist or antireductionist positions. For example, an antireductionist may say that since one higher-level property usually corresponds to a number of systems able to realize the property in question, reducing the higher-level property would result in a disjunction of many distinct systems. This position, for instance, is held by Fodor (1974, 108). On the other hand, a reductionist may object that, although type–type reduction might be hard to achieve due to the one-to-many relationship between the higher-level property and their distinct realizers, token–token reduction can still be made in the context of scientific explanation. That is, because under each specific circumstance at a given time, a higher-level property token corresponds only to a specific mechanism token. This position has, for example, been embraced by Weber (2005, 48).

I am sympathetic with both Fodor and Weber. Fodor is right to point out that reducing the higher-level property may result in a disjunction of many heterogeneous systems, while Weber is right to hold that token–token reduction can still be made in the context of scientific explanation. To resolve the dispute, however, I suggest that we divert our attention to the contexts in which scientific explanation occurs. After all, the dispute here concerns scientific explanation, and sticking to the MMRT itself does not get us off the ground. So, my view is that, on its own, the MMRT neither demonstrates nor refutes explanatory reductionism or antireductionism. Nevertheless, when allied with a view of scientific explanation (according to which explanation is by and large an *ontic-oriented epistemic* issue that must take into account contextual elements, extra information and cost-and-benefit considerations), we may find that the MMRT does cast some doubt on the plausibility of explanatory reductionism. In particular, the MMRT implies that in the context of scientific explanation, a higher-level phenomenon may be partially independent of (or insensitive to) its lower-level underpinnings, because changing the lower-level underpinnings (i.e., via different realizers) does not necessarily result in corresponding changes in the higher-level phenomenon. As such, an explanation of the phenomenon in question may just invoke coarse-grained causes at the higher-level without detailing the phenomenon's lower-level underpinnings. I discuss this issue in more detail in the next chapter.

2.6 Conclusion

This chapter showed that, contrary to Polger and Shapiro's claim, MR is not as difficult to find in the biological world as has been assumed. To establish this conclusion, I first examined Polger and Shapiro's criterion for MR, namely, their Official Recipe. This criterion is a worthy contribution to the MR debate because it offers us a useful and operable tool with which we can judge when there is a case of MR and when there is no such case. With this criterion at hand, and with the clarifications of the conception of the realization relationship, a multiple mechanistic realization thesis was developed. The MMRT is best understood as a complement to Polger and Shapiro's Recipe (rather than as a competitor). It is complementary because it attempts to integrate improvements and consensuses about the conception of realization that Polger and Shapiro's Recipe has not incorporated. The only disagreement between myself and Polger and Shapiro on this front concerns their assessment of the scope of MR. For Polger and Shapiro, cases of MR in the biological world are very rare. To disprove Polger and Shapiro's claim about the scope of MR, two examples (representative of two broad kinds of phenomena) were scrutinized in terms of the MMRT. The conclusion reached was that MR (MMR in particular) is much easier to find in the biological world than is acknowledged by Polger and Shapiro.

This chapter has also touched the problem that higher-level properties (states, patterns, functions, etc.) tolerate intra-kind variations. That is, members of a kind are not expected to manifest exactly the same (or identical) property. Polger and Shapiro's criterion admits this (recall the third and fourth conditions of their Official Recipe), and the MMRT also accepts this. The next chapter shows that it is a basic practice in science that scientists treat instances of a certain pattern of phenomenon realized by distinct underlying mechanisms as a single kind, even though they know that there are variations among these instances with respect to the pattern in question. In other words, in practice (e.g., in the context of scientific explanation), similarity—rather than identity—of a property or function is typically what we need in order to classify two instances as being of the same kind at the higher-level and the degree to which one instance is judged to be similar to another is decided by the practice itself.

This chapter also considered the implications of the MMRT for the reductionism debate in the philosophy of biology. I argued that, by itself, the MMRT takes no sides in this debate. In particular, the MMRT neither demonstrates nor refutes the explanatory reductionist position. I suggested that, to take issue with the explanatory reductionism debate, the MMRT must be allied with an account of scientific explanation, which is the task of the next chapter.

Notes

1 For instance, levels sometimes mean different levels of abstraction, sometimes mean different sizes of grain of descriptions, and still sometimes mean different spatial locations.
2 See, for example, Kim (1992, 1999), Bickle (2003, 2010, 2013), Bechtel and Mundale (1999), Sober (1999), Couch (2004), Klein (2008, 2013), and Polger (2008, 2009).
3 See Rosenberg (2001c, 2006), Gillett (2002, 2003, 2007, 2010), Aizawa (2007, 2009, 2013), Aizawa and Gillett (2009a, 2009b, 2011), Richardson (2008), Balari and Lorenzo (2014), and Piccinini and Maley (2014).
4 In fact, to claim that property *P* realizes property *Q* is too abstract and inaccurate. A more appropriate way of describing such cases will be introduced in Section 2.3.
5 Shapiro is not the first to have proposed a "causally relevant differences" criterion. Shoemaker makes a similar point in his causal theory of properties, stating that two properties are different when they contribute different powers to the individuals in which they are instantiated when under the same conditions (Shoemaker 1980; cf. Gillett 2010, 169). Kim's principle of causal individuation of kinds also asserts that "kinds in science are individuated on the basis of causal powers; that is, objects and events fall under a kind, or share a property, insofar as they have similar causal powers" (Kim 1992, 17).
6 With respect to the conception of function, they say that "for some entities—properties, states, kinds, objects—being that entity is a matter of *having* a certain function" (Polger and Shapiro 2016, 23; original emphasis).
7 Aizawa and Gillett distinguish three kinds of realization relationship, including linguistic realization in semantics, "abstract" realization in mathematics, and mechanistic realization that holds between the whole and its parts (2009b, 183).
8 "*Aggregate realization* is realization by constituents where the property of the whole is a simple sum of the properties of the parts.... The mass of a pile of sand is realized aggregatively by the masses of the individual grains" (Craver 2004, 968; author's emphasis).
9 The term *mechanistic* should be understood in the context of the neo-mechanistic framework introduced by Machamer, Darden and Craver, according to which "mechanisms are entities and activities organized such that they are productive of regular changes from start or set-up to finish or termination conditions" (2000, 3).
10 For different types of realization see Note 7.
11 *S's* and *B's* are properties at different levels defined by different sciences.
12 This is, in turn, because a property of a mechanism is not simply the sum of the properties of its components, for there are interactions among the components that also affect the realization of the property of the mechanism. I thank Patrick McGivern for helping me to clarify this point.
13 Traditionally we describe the relation between an individual and a property as one of instantiation, with the individual instantiating the property (Orilia and Swoyer 2016). Here, following Aizawa and Gillett (2009a, 2009b, 2011),

I describe the relation as one of realization, with the individual realizing the property. I think this usage is compatible with the traditional one, for they highlight different aspects of the relationship between an individual and a property: my usage stresses the particular way of organization of an individual's components that makes the individual have a certain property, while the traditional usage underscores the fact that it is an individual that has a property.

14 This example comes from Gillett (2002, 318-320).

15 In more recent work, Aizawa and Gillett (2009a, 2009b, 2011) make a correction to their conception of realization, according to which it is an individual, entity, or instance that realizes a certain property.

16 This definition draws heavily on Aizawa and Gillett's account (2009a, 2009b); however, this account differs in two substantial ways: (1) it makes clear that MR is a mechanistic relationship (i.e., a type of the compositional relationship), and (2) it attempts to integrate Polger and Shapiro's criterion for MR.

17 Note that properties and kinds are two different concepts. However, I sometimes use kinds to mean that something has a property *P* such that, due to having this property, it is classified into the kind *X*. I thank Patrick McGivern for alerting me to the difference between these two concepts.

18 There is another reason for concentrating on the second type of MMR: Polger and Shapiro's proposal of MR focuses on how a function or property can be realized in different ways, and the different ways are mainly associated with different types of organizations (or structures) rather than with different types of components organized in certain ways (Polger and Shapiro 2016). If we were to easily find cases of MMR2, then we would be able to respond to Polger and Shapiro's (2016, 111) challenge that it is relatively harder, if not entirely impossible, to find MR in the biological world than in the artificial world created by human beings.

19 The value of the index is achieved by calculating "the ratio of the nearest-neighbor's distance between the points in the data set and the distance between a randomly selected location in the field and the point from the data set closest to it" (Ruthazer and Stryker 1996, 7255; also see Hopkins and Skellam 1954).

20 See, for example, Martindale (1991), Anderson (1995), Wu and McLarty (2000), and Heaton (2015).

21 See Mostafavi et al. (2011, 399); also see Soffer and Vázquez (2005), Saramäki et al. (2007), and Gursoy et al. (2008).

22 I thank Arnaud Pocheville for suggesting this expression to me.

23 See Fung et al. (2011, 311–336).

24 An alternative measure of a network's global topological characteristic of clustering is *density*, corresponding simply to the ratio of existing to possible links in a network. For a discussion of these two measures, see Kaiser and Hilgetag (2004) and Kantarcı and Labatut (2013).

25 I do not show the values of the $C(k)$ for the three cortexes for two reasons: (a) the original paper of Sharma et al. (2000) only shows the values of the index of cell aggregation and (b) more important, I think the qualitative discussion is sufficient for my current purpose: the value of the $C(k)$ obtained from the rewired auditory cortex differs from the values obtained from the normal auditory and normal visual cortexes.

26 All enzymes are proteins except for a small group of catalytic RNA molecules (Nelson and Cox 2008, 184).

27 *Homologous* refers to genes, proteins, processes, functions, or even behavioral patterns and cognitive features that come from common ancestry (Brigandt and Griffiths 2007).

28 I thank Patrick McGivern for alerting me to the point that the examples discussed in this chapter concern not only the number of instances of MR but, more important, the variety of types of the instances as well.

29 Note that the stance of metaphysical reductionism has two versions: (a) a weaker version: token–token reduction that "each particular biological process (or token) is metaphysically identical to some particular physico-chemical process" (Brigandt and Love 2015) and (b) a stronger version: type–type reduction that "each type of biological process is identical to a type of physico-chemical process" (ibid.). Given that the weaker version is less controversial than the stronger one and becomes a default position nowadays among philosophers, throughout this book, whenever I refer to metaphysical reductionism, I mean the weaker version.

3 Explanation in Biology

Context Dependence, Extra Information, and Pragmatics

3.1 Introduction

The last chapter defended the mechanistic multiple realization thesis. As said there, since this thesis, in and of itself, remains neutral to the (epistemic) reductionism debate, no one expects it to offer any support to either side of the debate. However, that the thesis taken alone does not offer any direct support does not mean that it is useless. On the contrary, we will see in this chapter that when it is well placed within the context of scientific explanation and firmly connected to the fact that the higher-level is partially independent of the lower-level due to multiple realization, it can offer some help. So, the theme of this chapter is about explanation, together with its related philosophical stance, that is, explanatory reductionism (ER).

Although different versions of explanatory reductionism[1] can be found in the literature,[2] the key point is relatively simple: explanatory reductionism is the view that a higher-level feature (such as a mechanism, a generalization, etc.) can be further explained by a relatively lower-level feature (Brigandt and Love 2015) and that the explanatory significance is primarily, if not completely, located in the lower-level (Wimsatt 1974; Waters 1990, 1994, 2000, 2008; Sarkar 1992, 1998, 2001, 2002, 2005; Weber 2005).[3] As can be anticipated, the dispute surrounding reductionism versus antireductionism largely stems from different interpretations of the practice of reductive explanation. In particular, it stems from the fact that philosophers disagree over the extent to which the practice of reductive explanation is exercised in the biological sciences. For explanatory reductionists, reductive explanation is a dominant feature and is even further believed to be the only feature of significance. For opponents of this view, reductive explanation at best captures a fraction of biological practice. This chapter is devoted to showing that, for various reasons, the extent of explanatory reduction is not as widespread as reductionists envision.

As simple as ER may appear, crucial subtleties must be noted before proceeding. First, for proponents of this view, unlike traditional reductionism advanced by Nagel (1961) and Schaffner (1967, 1969, 1974,

DOI: 10.4324/9781003148029-3

1993), the currency of reduction is always cashed out by descriptions, empirical generalizations, laws, mechanisms, or even individual observation reports rather than theories (Sarkar 1992). So there is no such formal requirement (i.e., the condition of derivability) that the reduced theory be deduced from, with the help of bridge laws or correspondence rules, the reducing theory. There is also no such requirement (i.e., the condition of connectability) that the predicates of the reduced theory be systematically connected with, or defined by, their counterparts of the reducing theory. For, as the advocates of this view routinely put it, there is simply no general or universal theory in any subdiscipline of biology, for example, classic genetics, molecular genetics, population genetics, developmental biology, microbiology, physiology, ecologym, and so on, and what they have are at best fragments of a would-be theory or theories (Sarkar 1992).

Second, because of the first reason, reduction in the biological sciences normally proceeds in a piecemeal manner, as opposed to the systematic way required by traditional reductionism (Sarkar 1998). This leaves the room for not only granting the fact that explanatory reduction in biology is just being undertaken and thus far from being accomplished but also proffers the reason for taking the optimistic stance that there is no *a priori* reason to dismiss the possibility that it would be achieved one day (e.g., Rosenberg (2006) is such an optimist). Third, as indicated by the second point, advocates only concede that the desired reduction holds within subdisciplines of biology, for example, classic genetics and molecular genetics rather than between physics/chemistry and biology (Sarkar 1998; Waters 2008).[4] Fourth, adherents of this view, except for Waters (2008), have in their mind a "multiple-level organization" worldview best elaborated by Wimsatt (1974, 1976), according to which the living kingdom contains many levels of organization (e.g., molecules, cells, organelles, tissues, organisms, etc.) and an entity of one level, which itself constitutes a component of its upper level, is typically composed of entities of its lower-level. Hence, ER can be reformulated such that explanations at high levels such as cellular are inferior or at least incomplete and eventually can be further explained by features located at the level of molecules.

Although explanatory reductionism might have some merits, I think its scope and viability have been largely overstated. In this chapter, I argue that explanatory reduction captures at best one dimension of biological practice, but there are other dimensions in which a sweeping reductive strategy does not take hold. The argument of this chapter is proposed strictly along the *epistemic* line, in which by *epistemic*, I mean two interrelated matters. First, I adopt an (ontic-oriented) epistemic, as opposed to an *ontic*, conception of scientific explanation throughout the whole book, as explained in the next section. Second, I think both sides of the debate should at least concur that the plausibility (or implausibility) of explanatory reductionism should be best judged on the basis of actual scientific practice. So a couple of examples drawn from scientific practice are scrutinized throughout the chapter.

The layout of this chapter is as follows: Section 3.2 explains the (ontic-oriented) epistemic conception of scientific explanation. Section 3.3 considers the context argument leveled against ER. Section 3.4 considers the possibility that explanations in many areas of biology cannot be achieved without the supplement of knowledge from domains other than molecular biology. This kind of knowledge may include, but is not restricted to, spatial, structural, geometrical, topological, or mathematical information that cannot be simply subsumed into, nor reduced to, the domain of molecular biology. Section 3.5 scrutinizes the *pragmatic* problems for ER more broadly.

3.2 An Ontic-Oriented Epistemic Conception of Scientific Explanation

This chapter, and the whole book in general, are essentially about issues related to scientific explanation, so it is better to make sure what I mean by *explanation* from the very beginning. When it comes to scientific explanation, Wesley Salmon (1977, 1984, 1989) distinguishes three different conceptions. The first is the ontic conception, according to which explanations are physical entities in the real world, worldly things independent of human minds and objective features detached from human interests that "subsist in re and participate in the causal structure of the world" (Wright and Van Eck 2018, 999). Or, as Carl Craver (2007) puts it explicitly using his terminology "objective explanation", "objective explanations are not texts; they are full-bodied things. They are facts, not representations.... There is no question of objective explanations being 'right' or 'wrong', or 'good' or 'bad'. They just are" (27). The second is epistemic conception, according to which an explanation "could be described as an argument to the effect that the event-to-be-explained was to be expected by virtue of the explanatory facts" (Salmon 1984, 16). Viewing explanations as arguments should come as no surprise, given the traditional deductive-nomological model of scientific explanation developed by Hempel and Oppenheim (1948), according to which "to explain is to provide a sound deductive argument for some explanandum" (Weslake 2010, 274; we shall revisit this model in Chapter 4, Section 4.2). However, it is important to note that today's supporters of the epistemic conception have already updated their understanding in the way that they no longer regard explanations as arguments but rather as representations, descriptions, and models that deliver scientific knowledge. The last one is modal conception, according to which to explain is to show a relation of nomological necessity between the antecedent-conditions and the events-to-be-explained by virtue of some general laws.

The debate is usually cast between the ontic and epistemic conceptions when philosophers raise the question of which conception better makes sense of scientific practice. Proponents of the ontic conception (Craver 2007, 2014; Jenkins 2008; Strevens 2008; Illari and Williamson 2011) hold that the ontic conception fares much better, while supporters of the updated epistemic conception (Bechtel and Abrahamsen 2005; Bechtel 2008; Wright 2012, 2015; Sheredos 2016; Şerban 2017; Wright and Van Eck 2018) argue that the epistemic conception dovetails much better with scientific practice. Yet my purpose here is not to try to settle the debate but, rather, to lay out my understanding of scientific explanation and explain why I favor that understanding. Following Maria Kaiser (2015), I am sympathetic to an *ontic-oriented epistemic conception* of scientific explanation. According to this conception, explanations are not only just representations, descriptions, models, or any other representational means but are also the kind of means that aims to deliver scientific knowledge about the *real world*. Hence, explanations "are not the causes or parts of the causal structure of the world themselves, but rather descriptions or representations of these causes and partitions of the causal structure" (Kaiser 2015, 146). This conception is ontic-oriented simply because it insists that the explanatory power of scientific explanations does not come from their being able to be nomologically deduced from some general laws or being able to be logically expected from some antecedent conditions but rather from the fact that they track the actual causes or cause structures of the real world to some extent.

Some authors term this conception the *weak* interpretation of the ontic conception, in comparison with the *strong* interpretation that explanations are real entities in the world. Glennan (2002), for instance, takes such a stance:

> Causal-mechanical explanation exemplifies what Salmon calls the ontic conception of explanation. Explanations are not arguments, but are rather descriptions of features of a mind-independent reality—the causal structure of the world.
>
> (343)

However, I disagree that this is a weak interpretation of the ontic conception, for I think it is no longer ontic at all. As Wright and Van Eck (2018) have pointed out, those who think the ontic-oriented epistemic conception is a weak version of the ontic conception simply downplay the irreconcilable nature of the two conceptions, for "[t]he central divergence between proponents of [the epistemic conception] and [the ontic conception] concerns whether explanations are representations of entities in the world or the worldly entities so represented" (1001). Hence, they conclude that "there is no such middle ground to be had" (ibid., 1001). This is also what is in my mind. As long as we admit that explanations are representations, descriptions, models or any other representational means but not the entities (or relations) themselves in the real world so represented, it is not ontic. Therefore, an explanation is either ontic or epistemic but not both—it is certainly not in the way that one can be an ontic-oriented epistemic explanation as well as be an ontic explanation.

Given that the ontic and epistemic gulf is so deep and sharp, the next question is why I bolster one but not the other. The answer is simply that, as well documented in the literature, the ontic conception confronts—whereas the epistemic conception does not—serious problems when it comes to scientific practice. To begin with, if explanations are entities in the world, then it leaves no room for the fact that there are always competing, sometimes even conflicting, explanations for the same phenomenon of interest (Waskan 2006). But as a matter of fact, one can hardly deny that we, in practice, can always build different explanations to account for a given phenomenon of interest. By contrast, the epistemic conception does not have this sort of worry, because according to this conception, explanations are just *fallible* representational means about the world. Hence, this conception not only leaves room for the fact that there are always competing explanations for the same phenomenon of interest but also remains open to the possibility that some explanations might be much better than the others with respect to tracking the same target phenomenon. One potential reason for that possibility, for example, might be simply that some explanations track the target phenomenon much more closely or accurately than the others.

Second, as Bokulich (2016) argues, regarding explanations as entities in the world renders scientific investigation unnecessary, for explanations are already there no matter whether they are known or unknown. Yet, an advocate of the ontic conception might respond that even though explanations—just like rocks and bacteria—are always over there completely independent of human activities, it is still up to scientists to discover them so that scientific investigation is by no means unnecessary. Nonetheless, this move immediately reintroduces the serious problem discussed earlier: if it is still up to human beings to discover those explanations, then nothing can prevent the scenario from happening that numerous competing explanations regarding the same phenomenon of interest will always be "discovered" by scientists. Hence, as long as we are again faced with various competing explanations and to the extent that we somehow must make a choice among these competitors, it sounds implausible to assert that explanations are like rocks and bacteria over there completely independent of human activities. Furthermore, given the inevitable multiplicity of competing explanations regarding the same phenomenon of interest, the advocate of the ontic conception then owes us an answer about which explanation is *the* right explanation and why. But this all-too-important question puts them in a dilemma, for they can either go away and leave it unanswered or try to answer it but drop into the epistemic camp. Namely, if they ever try to answer this question, they can say nothing but that the right explanation is one that accurately tracks the causal structure of the world. However, this already falls into the epistemic conception of explanation. Of course, they can say something different, for instance, that it is the right explanation because it is just the right entity discovered. But this answer cannot even take off the ground, for it simply repeats the question rather than offers any hint (for there remains the question of why it is the right entity at all).

Third, as Târziu (2018) points out, if explanations are entities fully independent of human activities, then it makes no sense to talk about unexplanatory (or unimportant) components of explanations, as well as makes no sense to talk about different degrees of explanatory relevance. As he puts it, "[i]f explanations are 'full-bloodied things' ..., then either something is or is not part of an explanation (i.e. it is or it is not explanatorily relevant) and if it is, it cannot fail to be important" (2018, 410). In other words, the ontic conception can tolerate no degrees regarding explanations, be it the degree of explanatory power, the degree of importance or the degree of explanatory relevance. But it is common practice in science (as well as in our daily life) that we always deem some factors more explanatorily relevant than others. For instance, we all agree that an effect usually has many (perhaps innumerable) causes while a cause can often bring about numerous (perhaps even uncountable) effects so that an effect as simple as a match's lighting can be traced to a number of causal factors, for example, the presence of oxygen, some person's striking one side of the matchbox with the head of a match, the right angle and strength of the striking, the right dryness of the matchbox, the match, and so on. Yet, although acknowledging that all these factors are causally relevant (i.e., they all play a role in the matching's lighting), we usually think of the person's striking as the most causally relevant factor. Accordingly, we deem the shorthand explanation that "the person's striking causes the match's lighting" more appropriate than the "all-encompassing" explanation that "the person's striking, the presence of oxygen, the right angle and strength of the striking, and so on cause the match's lighting", with respect to the effect to be explained that the match lights.

The list of criticisms is not meant to be exhaustive, nor is either of the criticisms suggestive of a knock-down argument, but I hope the list does give the reader a clue that defending the ontic conception is not an easy task. It also becomes clear why I favor the ontic-oriented epistemic conception, a conception of scientific explanation that I will stand by throughout the whole book. Besides, one implication of upholding such a position will not showcase itself until we start to pay attention to the pragmatic dimension of scientific explanation in Section 3.5.

3.3 The Context Argument Revisited

It is a plain fact of molecular biology that the effect of a molecular entity (or process, pathway, mechanism, etc.) depends on the context in which it resides or occurs. Depending on different contextual factors (cellular, organelles, tissues, organismic, or even environmental), a molecular entity can give rise to many, sometimes even radically distinct, effects, a phenomenon often termed by philosophers as the *one-to-many*[5] relationship between the molecules and their upper-level entities (or processes, pathways, mechanisms, etc.; Hull 1972; Gilbert and Sarkar 2000; Laubichler and Wagner 2001; Burian 2004; Frost-Arnold 2004; Stotz 2006; Bechtel 2007).[6]

Aside from this context-dependence phenomenon, there is another way of raising the context argument against explanatory reductionism: the context independence of the higher-level phenomena from its lower-level underpinnings (e.g., changing the lower-level underpinnings from one to another does not necessarily change the higher-level phenomenon itself, and in this case, the lower-level underpinnings may be treated as a fixed context; Laubichler and Wagner 2001; Robert 2004). This section considers both the context-dependence and context-independence arguments in turn. Let us go to the first one.

3.3.1 The Context-Dependence Argument

The reason the context-dependence phenomenon constitutes an obstacle to explanatory reductionism is that knowing the molecular details underlying a phenomenon of interest is not sufficient, although important (or perhaps more strongly, necessary), to explain that phenomenon. To successfully explain that phenomenon, contextual information is also required. The idea of context dependence is expressed by Laubichler and Wagner (2001):

> Again, the question is not whether genes and molecules are important in development, but whether it is possible to assign well-defined functions to them in the developmental process without taking the specific functional context of their expression into account.
>
> (59)

Although Laubichler and Wagner's discussion concerns specifically the domain of developmental biology, their conclusion is obviously relevant to the problem of context dependence in general. Remember the core idea of ER mentioned in Section 3.1: a higher-level feature can be further explained by a relatively lower-level feature. Given the context-dependence argument, the dispute between reductionists and antireductionists now becomes whether the context itself can be fully described molecularly. In other words, the question boils down to whether contextual factors can also be explanatorily reduced to the molecular level.[7]

This question, however, is believed to pose no more a barrier for reductionists than the original one. Megan Delehanty argues that with a well-articulated notion of a mechanism the contextual factors can be properly accommodated within a reductive explanation framework (2005).[8] Her argument begins with an account of mechanism advanced by Machamer, Darden and Craver (2000), suggesting that investigating mechanisms enables us to understand the debate about reduction in biology. This is because mechanisms typically give us causal explanations of the phenomenon in question and their *directionality* character (i.e., bottom-up, a mechanism arises in virtue of its components and their activities) is intimately relevant to the directionality of explanation involved in the reduction (Delehanty 2005, 721). Given that the boundary of a mechanism varies depending on the different sets of questions considered, some mechanism that was regarded as background factors in the past can be incorporated into the mechanism under different considerations. This extension strategy can be further exercised so as to incorporate far more contextual factors such that finally complex networks or webs of interactions can be built (ibid., 722–723).

Delehanty's description of the problem-driven extension practice (or sometimes condensing) in science is undoubtedly illuminating, but the extension practice cannot really remove the obstacle that ER faces, because the context, as an unremovable backdrop, is always there. Even if a contextual factor X_1 is for some reason incorporated into the mechanism, there will be X_2 that falls outside that mechanism, and even if X_2 is included into the mechanism, there will still be X_3 that resides outside that mechanism, and so on; by the same token, even if a complex network of interactions is unambiguously delineated, this network still has its own boundary and context. One may reply that (1) in any given *token* mechanism to be explained, its boundary can be relatively stably fixed, and (2) an explanatory reductionist strategy is not to reduce everything to molecules—rather, it is only to reduce what needs to be reduced: contextual elements that are causally relevant to, or responsible for, generating a given mechanism. Therefore, although a new context must be introduced once we reduce those 'old' bits of contextual elements responsible for generating a given mechanism, within the new context everything is tidily molecularly described. Hence, first, there is no such infinite regress of incorporating contextual factors, and second, we only reduce what we need.

This reply, however, faces both an in-practice and an in-principle problem. First, consider the in-practice one: with respect to the contextual elements that are causally relevant to giving rise to a mechanism of interest, it is not always the case that scientists reduce them to the molecular level. Consider, for example, the research strategy employed by developmental genetics. In developmental genetics, scientists routinely hold cellular and organismal factors fixed as the background, and in doing so, "the causal influence of these non-genetic resources is simply not assessed by *this* research method" (Robert 2004; author's emphasis; cf. Brigandt and Love 2015). That is to say, in developmental genetics, given a certain question of pursuit, scientists simply hold fixed certain cellular and organismal factors *as if* they did not causally affect the phenomenon of interest, although they may be pretty sure that these factors sometimes do have causal influence, more or less, one way or another, on the phenomenon of interest. Moreover, even the success of developmental genetics research lends no support to the reductionist suggestion that we reduce the context to the molecular level, because the success, that is, "the discovery of genetic causes[,] occurs against a fixed organismal context" (Robert 2004; cf. Brigandt and Love 2015).

For the in-principle problem, consider the explanation of carcinogenesis. In explaining the phenomenon of carcinogenesis, the somatic mutation theory of cancer (SMT for short) has dominated for the past half century (Boveri 1914; Soto and Sonnenschein 2014; cf. Montévil and Pocheville 2017, 2). The theory says that

> [c]ancer is a cell-based disease driven by somatic DNA alterations which increase cell proliferation (Hanahan and Weinberg 2000). Accordingly, most carcinogens are assumed to be so in virtue of being mutagenic. At the core of carcinogenesis is the appearance of 'cancer cells'.... These cancer cells are assumed to be the product of several successive mutations (on oncogenes, tumor suppressor genes, DNA repair genes, etc.) which, supposedly, make these cells proliferate more, leading to their higher fitness (in the population genetics sense). Normal cells are assumed to be quiescent by default and to require 'signals' in order to proliferate. Cancer cells do not. As a result, cancer is assumed to be a (problematic) self-sustained cell proliferation.
>
> (Montévil and Pocheville 2017, 2–3)

As the explanation stands, the analysis proceeds at the cellular level, and the phenomenon of carcinogenesis is believed to be reductively explained in terms of mechanisms associated with cells. However, recent investigation has demonstrated that "cancer is essentially a developmental disease, occurring at the level of the tissue", and "carcinogenesis is understood as a disorganization of the morphogenetic field of the tissue" (Montévil and Pocheville 2017, 3). In particular, according to the tissue organization field theory (TOFT for short) proposed by Sonnenschein and Soto (1999),

> [h]ealthy tissues impose constraints on cell proliferation (via mechanical forces, chemical inhibitors, etc.). However, a disruption of tissue organization can release those constraints, resulting in cell proliferation with variation and motility, and in further disorganization of the tissue. Carcinogens are assumed to be so in virtue of altering the tissue architecture (e.g. asbestos), or of interfering with normal development (e.g. endocrine disruptors).
>
> (Montévil and Pocheville 2017, 3)

In other words, to explain "cancer cells" we must go one level higher to the tissue in which the cells find themselves. As an indication, consider epithelial cancer. This cancer "involves reciprocal interactions between the two main parts of the considered tissue, the epithelium which typically proliferates abnormally, and the stroma which surrounds the epithelium" (ibid., 3). More important, "being a 'cancer cell' is not a genuine property of the cell: 'cancer cells' do not acquire new competences and they can be normalized if placed in an appropriate tissue" (ibid., 3–4). Hence, a cell can be viewed as dependent on the tissue in which the cell finds itself, because the tissue constrains how the cell behaves. When the tissue is normal, it is typically considered as constant or fixed, that is, considered as the context. When, on the other hand, the tissue is abnormal, we start to pay attention to the tissue (i.e., the context). However, paying attention to the tissue does not necessarily result in reducing the tissue to its cell components, nor can we do so, as the carcinogenesis case has shown.

The take-home message from the example described earlier is that, in explaining a phenomenon of interest, it is not always the case that we are able to explain the phenomenon if we reduce the context involved to its lower-level details, for example, reducing the tissue to its cellular details. Rather, the actual scenario is that, in order to properly explain a certain phenomenon, we sometimes need both the bottom-up (reductionist) and the top-down (non-reductionist) perspectives, for example, both from cell to tissue and from tissue to cell in the carcinogenesis case (note that the case does not say that we do not need to consider the cellular level; it only says that we must also consider the tissue level; ibid., 4). The problem is not that taking a top-down perspective helps us *better* explain a phenomenon of interest but rather that we *cannot* explain the phenomenon without taking into consideration the top-down perspective.[9] That is, it is not a problem of degree but an all-or-nothing issue. Thus, it is in this sense that I think the reductionist encounters an in-principle obstacle when attempting to reduce the context involved in a phenomenon to its lower-level underpinnings.

The in-principle contextual dependence problem can be approached through a different but related angle: contextual elements are usually incorporated into *ceteris paribus* clauses: *other things being equal*, then such and such. For instance, it often takes the following form: if other things were kept constant or fixed, then doing this would result in that, and so on. Sometimes scientists make the clauses explicit, but most of the time scientists treat them implicitly—it is part of the practice of a scientific community, in which members of the community state something clearly but leave other things to the background. A cursory look at science can show this practice: in scientific modeling, modelers routinely group all those unmodeled or unknown causes pointing into a variable X into a single error variable (e.g., ε_i) as if there exists such a single variable—even though they know that there might exist a set of heterogeneous causes that affect the variable X (Chapter 5 gives an example about the modeling practice wherein a couple of error variables are involved). The practice is rooted in a deep ground: one can never specify all those factors that may be causally relevant to bringing about the phenomenon of interest (see Lange [1993]; Earman and Roberts [1999]; Earman, Roberts and Smith (2002) for relevant discussion about ceteris paribus clauses). Therefore, it is not simply that we do not need to specify all these causes but rather that we in principle cannot specify all of them. If it is impossible to specify these factors in principle, then the hope of reducing them becomes dim. The next chapter says more about this point when it comes to discussing ceteris paribus laws.

In short, it is not always the case in practice that scientists reduce contextual factors to the molecular level (even if they are known to be causally relevant), nor is it the case that in principle they can be easily reduced. The more common practice is that scientists keep them fixed as a normal background.[10] In doing so, they have good reasons. Admittedly,

> we are limited beings, and most of the systems we want to understand are too complex in their natural state; thus we abstract from them what seem to be the most important or the most easily manipulated variables in order to generate a manageable representation of their workings.
>
> (Robert 2004, 3)

More importantly, the most common simplifying strategy

> is to simplify the context of a system under study. If we want to learn about *intrasystemic* causal factors—that is, if we want to learn about what's going on inside a particular system—we build a model or design an experiment wherein the context of the system is simplified rather than the system itself.
>
> (ibid., 3)

This strategy is best exemplified by exploring the role played by genes in development as mentioned earlier, wherein one routinely varies genetic factors against an assumed constant background.

In sum, we found that scientists often focus on something, keeping other things in the background. They do not always exercise the sweeping reductive practice. In doing so, they have both practical and in-principle grounds. If our aim is to understand scientific practice, then we cannot dismiss this.[11]

3.3.2 *The Context-Independence Argument*

Now let us turn to another side of the coin, the context-independence argument.[12] The foregoing argument concerns cases wherein scientists treat higher-level contextual elements (e.g., cellular and organismal) as fixed when exploring molecular-level mechanisms. But the opposite is also true in certain areas of investigation, say, in developmental biology, where scientists treat the molecular-level mechanisms as fixed in the effort to explore higher-level causes. Consider a classical example from the developmental evolution literature (Alberch and Gale 1983, 1985). Scientists have wondered why species of frogs and salamanders that have lost phalanges and even whole toes have done so in a very ordered manner; that is, certain skeletal elements are lost prior to others (Alberch and Gale 1985, 8). In particular, the question under investigation is why salamanders always lose their digits 5 and 4 first when they reduce their digit number in evolution, whereas frogs first lose digit 1 followed by digit 5 then digit 2. Alberch and Gale (1985) claim that

> [i]ndependent of the specific pattern formation mechanism, all dynamical models proposed to explain limb morphogenesis exhibit the property that the resultant pattern is to some degree dependent on the size of the embryonic field at the time of pattern specification. (16)

That is, the pattern is produced by simply reducing the number of primordial cells prior to skeletogenesis: when the number of primordial cells of the limb bud is within a certain range, the limb will develop a normal structure, whereas when the number of cells is reduced below a certain threshold, a discrete change will occur in terms of loss of phalanges or even complete toes (ibid., 17). Moreover, the sequence and pattern of losses will not be random but highly structured: the last digit to be formed in ontogeny is the one to be lost first in phylogeny. In particular, in frogs, the last two digits that are formed are the digits 5 and 1 so that they are most likely to be lost, whereas in salamanders these digits are 4 and 5 so that they are most likely to be lost (ibid., 17).

In concluding their research, Alberch and Gale (1985) state that "the reduction in number of cells prior to the process of pattern formation *causes* a respecification of the number and spatial organization of skeletal elements" (17; my emphasis). This means that any change that has the effect of reducing the number of primordial cells below a certain threshold prior to skeletogenesis will have the effect of changing the developmental pattern of the limbs.

The most straightforward message we get from this example is that the explanation of the pattern formation does not invoke anything genetic or molecular. It only cites the reducing of primordial cells as a causal (explanatory) factor. This, of course, does not amount to saying that the process of pattern formation is totally free of its underlying molecular underpinning. Surely it has a molecular underpinning. But the fact is that, as Alberch and Gale put it clearly at the end of their essay, the process of pattern formation is partly independent of its genetic underpinning in the sense that changing its genetic underpinning from one to another does not necessarily change the pattern formation itself:

> Our basic premise is that morphological change does not necessarily correspond to a specific genetic change, i.e., changes in many genes affecting different morphogenetic parameters may result in the same morphological change.... In both cases, if the cell number is small enough, the developmental interactions generating a global pattern will produce a very specific, and invariant, skeletal transformation, in spite of the fact that different underlying parameters were altered genetically. (ibid., 19)

Because of this partial independence of the pattern formation from its lower-level molecular underpinning, the latter in this case is simply treated as a fixed background. In this case, it can also be said that the lower-level explanation based on genes or molecules is simply *screened off* by the higher-level explanation which only involves the reducing of primordial cells.[13] Rather than an exceptional case, however, more recent studies have shown that many interaction patterns (and metabolic pathways) are conserved across a number of different biological taxa (see, e.g., Sharan *et al.* 2005; Solnica-Krezel 2005; Koyutürk *et al.* 2006; Peregrín-Alvarez *et al.* 2009), indicating that it is common that higher-level patterns are in part independent of their underlying molecular underpinnings. One implication for my purposes here is that, insofar as the why-question is raised at the pattern formation level in development biology, the partial independence of the pattern formation from its lower-level underpinnings dictates a level of analysis at which a reductive explanation is not required. If one does feel the need to pursue further questions such as how a particular pattern is formed or how different molecular underpinnings can give rise to the same pattern, then she simply shifts our questions of concern, namely our explananda.

The example discussed so far reminds us of the multiple mechanistic realization thesis discussed in Chapter 2. That thesis can be interpreted as a foundation for raising the independence thesis discussed in this chapter because it states that many different lower-level underpinnings can realize the same higher-level property (or relation or status). In the context of scientific explanation, this entails that the higher-level property is thus partly independent of any of its lower-level underpinnings because changing the lower-level underpinnings from one to another does not necessarily change the higher-level property. Hence, an explanation operating at the higher-level does not necessarily need to appeal to the lower-level happenings. In other words, to use Salmon's phrase, because of this multiple realization relationship, the lower-level underpinnings are *screened off* by their higher-level property (Salmon 1971; we will revisit this term in the following sections). Chapter 2 said that the multiple realization thesis by itself does not take sides in the reductionism versus antireductionism debate. Now we see that when it is reapproached from the perspective scientific explanation, the thesis can be employed to lend support to one side of the debate, namely, the explanatory antireductionist side.

3.3.3 Summary

Together, the two arguments above have shown that the strategy of holding certain elements fixed as contexts runs in both ways: the higher-level elements are routinely treated as constant when exploring the lower-level elements, and the lower-level elements are usually treated as constant when exploring the higher-level elements. In the former case, the lower-level elements are dependent on the higher-level contextual elements, while in the latter, the higher-level elements are partly independent of their lower-level contextual elements. In either case, a one-size-fits-all reductionist strategy does not arise. For the former, it is not a *bona fide* reductionist strategy because contextual elements that may be causally relevant to generating a phenomenon of interest are not fully specified or reduced, nor can they be fully specified or reduced; for the latter, it is not a genuine reductionist strategy because the explanation operates at a level of analysis which is relatively independent of its underlying (molecular) happenings.

3.4 The Extra-Information Argument

Another line of argument against explanatory reductionism goes roughly that even if higher-level explanations/phenomena and their contextual factors can be fully molecularly spelled out (which is not true as discussed in the last section), there is a further handicap for reductionists: to understand a system, in addition to molecular information, many explanations, especially in developmental biology and ecology, must invoke spatial, structural, topological, geometrical or mathematical information in order to adequately answer the problem in question (Jackson and Pettit 1990; Sarkar 1998; Keller 1999; Frost-Arnold 2004; Batterman 2010; Huneman 2010, 2018; Dupré 2013; Woodward 2013b; Darrason 2018; Felline 2018; Kostić 2018; Rathkopf 2018). In other words, molecular knowledge may be necessary, but it is by no means sufficient.

To show this, consider an example in which geometrical information plays a central role in explaining phenomena. This example comes from ecology and was first introduced by Philippe Huneman (2010). Suppose we have two populations of bacteria: one with all individuals being fittest and another with most individuals being very close in fitness (see Figure 3.1). As represented in Figure 3.1, the distributions of individuals of these two populations are different: one shows a peak shape while another exhibits a flat shape. Now we put these two populations in competition and see which one will win over another. It turns out that the population with the flat shape ("the flattest" hereafter) will always win over the population with the peak shape ("the fittest" hereafter) when the mutation rate is high.

We wonder why this happens. In Figure 3.1, population *A* has a peak shape while *B* has a flat shape. More specifically, we say that these two shapes possess two different geometrical characteristics: *flatness* and

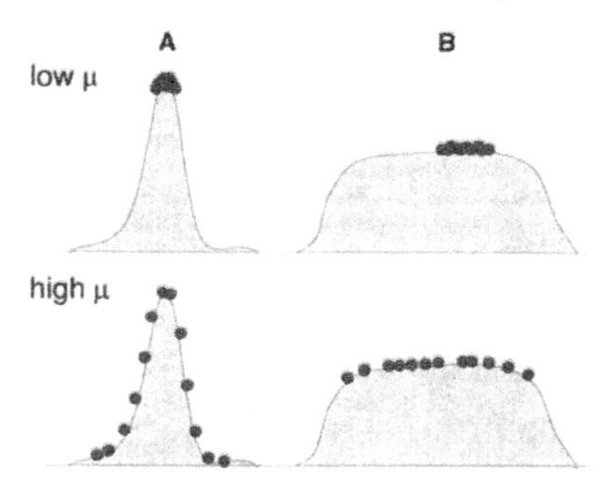

Figure 3.1 Survival of the flattest. μ is the mutation rate; vertical line is fitness. (Wilke *et al.* 2001; cf. Huneman 2010, 216). Figure used with permission.

sharpness. Given their different geometrical characteristics, the explanation simply runs as follows: the flattest always wins over the fittest when the mutation rate is high because the flattest has the geometrical characteristic of flatness whereas the fittest has the geometrical characteristic of sharpness; the reason why flatness and sharpness matter to our explanation is that, for the flattest, "a mutation has many chances to still be on the plateau", and, for the fittest, "a random mutation has many chances to occur in the steep of the fitness peak, and to decrease the fitness" (Huneman 2010, 215). In other words, the geometrical characteristics of the flattest and fittest guarantee that the fittest has a higher chance of decreasing the fitness than the flattest when the mutation rate is high. So we obtain a geometrical explanation, one in which we explain a phenomenon of a system in terms of the geometrical characteristics the system has.[14]

One may, however, ask why we should subscribe to the claim that geometrical explanations are explanatory at all. Fortunately, this question can be readily answered by appealing to James Woodward's interventionist account of scientific explanation. First, geometrical explanations, like mechanistic explanations, can help locate difference-making factors for their explananda, showing dependency relationships between these explananda and the factors cited as explanantia (Woodward 2013b, 63).[15] In the populations of bacteria, for example, if we were to modify the shape feature of population *A* to such an extent that it has the same shape feature as *B* (i.e., flatness), then we would be able to predict that neither *A* nor *B* could always outcompete the other when the mutation rate is high. Second, these dependency relationships are stable or robust enough such that they continue to hold under a range of changes in other background conditions (ibid., 63). For example, even if we were to substitute the population *A* with a distinct population *C* that has the same shape character as *A*, we would still be able to predict that should *C* and *B* be put in competition, *B* would always win over *C* when the mutation rate is high.[16]

Given that they are explanatory, a further question can be raised by a reductionist: Can we reduce these geometrical explanations to other lower-level explanations, say, mechanistic explanations? The answer is clearly no. First, and most important, geometrical properties can be partially independent[17] of their underlying processes (or events, states, activities, mechanisms, etc.) even if these underlying processes are causally responsible for bringing about the phenomenon under consideration (Huneman 2010, 218). This can be shown by slightly revising the bacteria example.[18] Suppose there are subpopulations B_1, B_2, and B_3 sharing the same geometrical property of flatness T_i, although each having distinct underlying causal processes within it. For example, members in B_1 tend to live closely, members in B_2 tend to live dispersively and members in B_3 tend to live neither too closely nor too dispersively. Besides, there is a population of bacteria *A* that possesses the geometrical property of sharpness T_j. Now let us put B_1 and *A*, B_2 and *A*, and B_3 and *A* in competition, respectively, and ask what the resulting outcome would be. Not surprisingly, it can be anticipated that all B_1, B_2, and B_3 will triumph over *A* when the mutation rate is high. The explanation for this remains the same as before: for *A*, "a random mutation has many chances to occur in the steep of the fitness peak, and to decrease the fitness", whereas for B_1, B_2, and B_3 "a random mutation has many chances to still be on the plateau" and thus to maintain the fitness of each (Huneman 2010, 215). This revised example shows that differences in underlying causal processes do not necessarily make a difference to the resultant overall upshot (i.e., populations with T_i always triumph over populations with T_j when the mutation rate is high), suggesting that explaining the phenomenon of interest can be exercised at the geometrical properties level without invoking the phenomenon's underlying causal happenings (Montoya *et al.* 2006; cf. Huneman 2010, 219).[19]

Second, even granted, for the reductionists' sake, that we can reduce geometrical explanations to mechanistic ones, it comes at a considerable price. Geometrical explanations, as a species of noncausal explanations (in particular, a species of mathematical explanations), characteristically possess a kind of *modal power* that their mechanistic (i.e., causal) counterparts lack. That is, they show not only that something *did happen* in such and such a way but also that what did happen *had to happen* in such and such a way.[20] In our example, a mechanistic explanation might say that due to such and such underlying causal factors (e.g., causal factors associated with the tendency to live closely), population *B*, *as a matter of fact*, triumphed over *A* when the mutation rate is high. By contrast, a geometrical explanation would say that no matter what the specific underlying causal factors would be, say, be the one associated with the tendency to live closely, a population with the geometrical property of flatness would *always* win over a population with the property of sharpness when the mutation rate is high. Therefore, insofar as scientific explanations concern not only the happening generated by a set of very specific underlying causal factors but also the more general phenomenon that can be brought about by many different underlying causal underpinnings, reducing the general explanation (i.e., the geometrical explanation) to the specific one (i.e., the mechanistic one) inevitably leads to the loss of a great deal of explanatorily important information.[21] To be sure, sometimes we might be only interested in the very specific happening. But even so, it remains the case that at other times we might be interested in the more general one. If this is the case, then a sweeping reductionist suggestion seems unwarranted.

It seems that in cases such as the populations of bacteria considered above, geometrical explanations fare very well, leaving reductive-oriented explanations unemployed. This sounds like good news for non-reductionists, although a more reasonable claim would be that geometrical (or, more generally, noncausal) and mechanistic explanations sometimes work hand in hand, without one completely superseding another (Andersen 2018; Darrason 2018; Huneman 2018). To show this, suffice here to briefly consider the explanation of the evolution of cooperation in which another type of extra information is invoked, namely topological information.[22] Cooperation involves an interaction between individuals that benefits the recipient but not necessarily the donor (Sachs *et al.* 2004). A hypothesis for this is that if cooperators are interacting with individuals that are also cooperators, then cooperation is likely to evolve (West *et al.* 2007; cf. Huneman 2010, 227). The theory of natural selection provides part of the explanatory story, which displays the exact mechanisms in which individuals interact with one another. So a mechanistic story has been given. However, topological properties of the network of cooperators are also at play: the topology of networks can also guarantee that cooperators are more likely to meet other cooperators (Nowak 2006; cf. Huneman 2010, 227; also see Van Baalen and Rand 1998). The connection between these two types of explanations is that the topological explanation (or, more generally, the extra information) constrains what kinds of cooperative interactions are possible and thus "provides the context within which natural selection yields different cooperative outcome" (Huneman 2010, 227).

Other cases of explanation invoking *extra information* (i.e., information that is complementary to, or cannot be fully exhausted by, the underlying causal details) to explain a phenomenon can also be found in many subdisciplines of biology. For instance, Sober's (1983) *equilibrium explanation*, Sarkar's (1998, 173) caveats about the *topology of the network*, Keller's (1999) *positional information* and Frost-Arnold's (2004) *spatial organization*, to list just a few. Nevertheless, examples elaborated in this section are sufficient for my current purposes: many explanations in a number of subdisciplines of biology must invoke spatial, structural, geometrical, topological or mathematical information in order to accomplish the job of explaining biological phenomena, information that is usually not located at the lower-level (i.e., molecular). Lower-level explanations, usually taking the form of mechanistic explanations, if recruited, at best work hand in hand with the *extra information.*

3.5 Pragmatic Problems

Section 3.3 discussed one particular facet of pragmatics of scientific explanation, that is, context sensitivity.[23] This section intends to reinforce the points resulted from the context-sensitivity argument by considering the pragmatic dimension of scientific explanation more broadly.

In one sense, the more serious problem of explanatory reductionism is not that because of the extreme complexity of molecular details or our computational imperfection, reduction in practice is impossible. Rather, there is a problem of why we should always privilege lower-level explanations over higher-level ones and thus take seriously the suggestion that higher-level explanations (plus their contextual factors) be explained by, reduced to or at least deepened by lower-level explanations. Looking closely, this "privileging-lower-level-explanations" suggestion appears as a philosophical fancy, detached from any specific context in which a given scientific explanation demand arises. Worse still, a corresponding methodological suggestion creeps in: in any case and at all times, scientists should pursue the lower-level explanations and discard the less lower-level explanations if the former are available or at least forthcoming.

However, even a cursory look at scientific practice reveals that there is no better explanation *simpliciter*; explanations are characteristically advanced for the purpose of answering specific questions, and thereby, the notion of better explanations can only be legitimately understood relative to specific questions being asked in specific contexts. Recall in Section 3.2 I said that I favor an ontic-oriented epistemic conception of scientific explanation, according to which an explanation is a human-made representation, description, model or any other form of representational means that aims at tracking the features (causal or noncausal) of the world. No doubt, this conception has a built-in ontic dimension, but it is also clear that the conception has an inherent epistemic dimension. That is, constructing explanations is a part, perhaps one of a few most crucial parts, of human activities that strives to make sense of their surrounding world. So, as long as this human activity goes, it must make sense of the world in the way that its audience, the explanation-seeker, understands so. This makes explicit one feature of explanation: it is a kind of *communicative act*, to borrow a term from Potochnik (2017, 124; see also Wright 2012). Therefore, explanation is construed as a dual-dimension human activity, an activity that needs not only align with the world in the right way but also align with the explanation-seeker in the right way. Furthermore, these two dimensions are not working independently, but rather, they are intimately interacting in a way that the epistemic dimension helps shape the ontic dimension, just as Potochnik (2017) says:

> [An] explanation's audience influences what should be represented and how it should be represented to generate a satisfactory explanation in any individual instance of explaining. One element of this influence is due to the audience's specific interests, that is, what they seek to understand. The combination of what and how an explanation represents, together with what the audience seeks to understand, determines what dependencies are featured in an explanation.... An explanation's audience thus helps determine the nature of the explanatory facts, that is, the ontic explanation.
>
> (128)

The interplay between the ontic and epistemic dimensions of explanation can be vividly illustrated with the toy example discussed in Section 3.2: a match's lighting. As I said there, we live in an extremely complicated world in which each effect might have an unlimited number of causes while a cause might generate innumerable effects as well. So, even for an event as simple as a match's lighting, there might be a huge number of causes that underlie that event's happening, for example, the presence of oxygen, some person's striking one side of the matchbox with the head of a match, the right angle and strength of the striking, the right dryness of the matchbox, and the match, to name just a few. Worse still, for each of these causally relevant factors, we can find an overwhelmingly detailed micro-level, relating to infinitesimal particles, for example, story to tell. Taking all these together, then, we might find ourselves in so awkward a situation that the project of providing an explanation cannot even take off the ground. However, we in reality do have a ready explanation within reach, one that simply says that some person's striking makes the match light. What is happening here is just this: First, we do not think the underlying micro-level stories are relevant to our current context; second, we take all the causally relevant factors but some person's striking as the background conditions of the event to be explained. So, ontologically speaking, all these factors are causally relevant (or are causally on par if you like), but we, due to contextual or pragmatic reasons, automatically shrink the ontic space in our mind to such an extent that we only single out the person's striking as relevant. Of course, we might single out some other factors as causally most relevant in some other contexts, but the general point remains the same: we usually merely pick out a particular (often small) proportion of the whole ontic space as causally most relevant to answering our current explanatory question at hand. Notice that the fact that the contextual or pragmatic considerations come after the ontic space is (at least partially) set up in this example does not mean that the former always come after the latter or are always less important than the latter. Rather, it might well be the case that contextual/pragmatic considerations sometimes come before the ontic space has even been explored so that the former essentially guide the exploration of the latter. For example, at the onset of exploring some new research area or question, our interests or concerns might push us to put some particular aspects of the system to be investigated in the limelight, even though we have not yet known the system too much.

All these elements originating from the audience, the explanation-seeker of explanation, are usually summarized under the umbrella term *pragmatics* of scientific explanation, which includes the audience's interests, goals, and concerns; the specific explanatory question to be answered; the specific context in which this question arises; the resources and knowledge available to the audience; and so on. This understanding of "pragmatics" of explanation is not alike that in linguistics, where *pragmatics* is related to how the meaning of a word or sentence is influenced by the context in which this word or sentence is expressed. From the perspective of this dual-dimension framework, we therefore can see that the reductionist overplays the ontic dimension of explanation while downplays its epistemic dimension, requiring that we always favor the lowest-level explanation(s) over the other candidate explanations. We now know that this requirement is unwarranted, for the "best" explanation usually varies from context to context. This means that the same explanatory question can have distinct, sometimes even inconsistent, explanations depending on disparate contexts of raising the question. This is plainly true in scientific modeling practice, in which, with respect to the same target system, a set of related but incompatible models (due to incompatible idealizing assumptions, approximations and/or simplifications) can be constructed in order to answer explanatory questions about the nature and causal structures that bring about the phenomenon under consideration (Weisberg 2013, 103). And even when these models are compatible, there can hardly be a single best model, because the selection of models typically involves trade-offs among various model properties depending on different goals of representations, such as accuracy, precision, and generality (Levins 1966; Weisberg 2006, 2013; Matthewson and Weisberg 2009). Some models may be more accurate than others, some more precise and still others more general, but there is no guarantee that a more accurate or more precise or more general model is always one that is described molecularly. Thus, simply asking which one is the best, without looking to the context in which the explanatory question arises, apparently misses the point in these cases. Moreover, addressing complex biological problems routinely calls for a heterogeneous combination of disciplines and theoretical approaches, among which more fundamental explanations are judged on a case-by-case basis (Craver 2007; Love 2008; Brigandt 2010).

In short, though ontic-oriented, providing explanations is inherently pragmatically driven. This idea can be found in seminal work such as Levins (1966), Salmon (1971, 1984), Wimsatt (1974), and Putnam (1975) and in more recent work such as Love (2008), Strevens (2008), Brigandt

(2010), Wright (2012), Kaiser (2015), and Potochnik (2017), to mention just a few. To give a taste of how pragmatic considerations percolate scientific explanation, let me start with Putnam's toy example concerning "peg-and-hole" and Sober's objection to it:

> Suppose we have a very simple physical system—a board in which there are two holes, a circle one inch in diameter and a square one inch high, and a cubical peg one-sixteenth of an inch less than one inch high. We have the following very simple fact to explain: *the peg passes through the square hole, and it does not pass through the round hole.*
>
> (Putnam 1975, 295; author's emphasis)

To account for this, both a lower-level explanation making an appeal to fundamental physics (e.g., quantum physics) and a higher-level explanation invoking only structural or geometrical facts can be given. Because the peg is a rigid lattice of atoms (so is the board), the lower-level explanation might involve a story about this lattice, its electrical potential energy, and so on; moreover, to complete the explanation, one may also need a story about all possible trajectories of the peg that are related with how the peg could pass through the square hole (Putnam 1975, 295). On the other hand, a different, simpler story can be told:

> The explanation is that the board is rigid, the peg is rigid, and as a matter of geometrical fact, the round hole is smaller than the peg, the square hole is bigger than the cross-section of the peg. The peg passes through the hole that is large enough to take its cross-section, and does not pass through the hole that is too small to take its cross-section.
>
> (ibid., 296)

Although it is true that the explanation involving particle mechanics is located at a compositionally lower-level than the geometrical one—few would deny this—it is also plainly true that the lower-level explanation is irrelevant. The geometrical explanation, although as simple as it appears, is good enough to answer the specific question being given. The lower-level explanation is "screened off" by the higher-level one, a phenomenon noticed by Salmon (1971). Of course, if a different question is raised, say, "Does the fact that the peg passes through the square but not the round hole have anything to do with the peg's underlying constituents?" then this time the geometrical explanation may be screened off by the lower-level one. Sober, however, disagrees with this verdict. He claims that "an ideally complete scientific explanation of a singular occurrence … would include the macro-story, the micro-story, and an account of how these two are connected" (Sober, 1999, 550). Furthermore, when it comes to the occasion of choosing one explanation among many, "it is a matter of taste whether we prefer the macro- or the micro-explanation", because "there is no objective reason to prefer the unified over the disunified explanation" (ibid., 551).[24]

I do not fully agree with Sober. To begin with, it seems unclear what the "ideally complete scientific explanation" (ICSE) consists of.[25] Let me dwell on the notion of an ICSE for a moment. In his 1999 paper, Sober refers to Hempel (1965) when introducing the notion of the ICSE, a notion that has been further developed by Railton (1981). Railton (1981) claims that there exists an ideal text for an explanation that takes the following form:

> An inter-connected series of law-based accounts of all the nodes and links in the causal network culminating in the explanandum, complete with a fully detailed description of the causal mechanisms involved and theoretical derivations of all the covering laws involved. This full-blown causal account would extend, via various relations of reduction and supervenience, to all levels of analysis, i.e., the ideal text would be closed under relations of causal dependence, reduction, and supervenience.... Such an ideal ... text would be infinite if time were without beginning or infinitely divisible, and plainly there is no question of ever setting such an ideal text down on paper.
>
> (247)

Yet, Railton admits that even "a whole range of less-than-ideal proffered explanations could more or less successfully convey information about such an ideal text and so be more or less successful explanation" (ibid., 247). In particular, "explanations single out or shed light on a particular part of some ideal explanatory text, which contains all information that is explanatorily relevant to the explanandum phenomenon" (ibid., 240; cf. Kaiser 2015, 167). For my current purposes, suffice to note that even for Railton, the ideal explanatory text serves only as one extreme of a continuum (and the nonexplanations serve as another extreme), within which genuine explanations reside by virtue of possessing part of the information of the ideal text (Railton 1981, 241). I think this is also what Sober really has in mind when invoking ICSE rather than the idea that an adequate explanation is one that contains *all* information of the ideal text.

So the remaining problem is how to single out the relevant part of the ideal text, and here is the place where my position diverges from Sober's. Recall that Sober (1999) holds "it is a matter of taste whether we prefer the macro- or the micro-explanation" in that "there is no objective reason to prefer the unified over the disunified explanation" and "science has room for both lumpers and splitters" (551). Nevertheless, though I agree that there is no good explanation *per se* and that "science has room for both lumpers and splitters" (ibid., 551), some explanations are relatively better than others with respect to a certain question being raised in a given context. In particular, there exist various practical (or pragmatic) grounds on which to single out one (or a few) explanation(s) over others. For example, one such pragmatic ground concerns the question being asked and the evidence (or data) we have with respect to answering that question.

To show such a pragmatic ground, let me start with an assumption often taken by reductionists: to appropriately explain a phenomenon of interest we need to collect as much information or data as possible, and lower-level explanations always convey more such information (e.g., Rosenberg 2006). Admittedly, we do need to collect as much information as possible because too little data might incur bias in estimation and prediction (Otsuka, 2021, 19). This, however, does not entail that adding more information (e.g., information about one system's lower-level details) always results in a better explanation; on the contrary, it sometimes harms the explanations. Statistically speaking, as indicated by the Akaike Information Criterion (Akaike 1974), there exists a trade-off between "containing too much unnecessary information" and "containing too little necessary information". The former situation corresponds to a problem called *overfitting*, where a model (or an explanation) contains more terms (or parameters) than needed or uses much more complicated approaches than needed, a problem that renders the model (or explanation) less plausible (Hawkins 2004, 1); the latter corresponds to bias, as said earlier. As a consequence, a good explanation is not one that simply contains as much information as possible but one that also strikes a good balance between overfitting and bias.

The trade-off, however, should not come as a surprise, if one remembers another pragmatic ground on which better explanations can be singled out, namely Wimsatt's *cost-and-benefit* consideration. Partly along the lines of Salmon (1971), Wimsatt deploys the notion of *effective screening off* to stress the relevance of cost-and-benefit analysis in evaluating scientific explanations. Wimsatt (1974) argues that because "most entities most probably interact most strongly with (and most phenomena are most probably explained in terms of) other entities and phenomena at the same level" (689),[26] there is an order of priorities in the search for an explanation in which appealing to the same level of entities or phenomena for explaining incurs less cost and more benefit and thus has the first priority (ibid., 689). Furthermore, "when a macro-regularity has relatively few exceptions, redescribing a phenomenon that meets the macro-regularity in terms of an exact micro-regularity provides no (or negligibly) further explanation" (ibid., 690). So in these cases, the lower-level explanations, when taking into consideration the cost-and-benefit balance, are simply "effectively screened off" by the higher-level explanations. The situation, however, is different when anomalies emerge or when exceptions increase to a certain extent. This is a situation in which the lower-level cannot be effectively screened off by the higher-level, and therefore, the priority should be shifted to searching for lower-level explanations that can account for the anomalies or exceptions (ibid., 690). In short, Wimsatt's cost-and-benefit consideration provides a very general analysis framework for practically choosing better explanations.

Given these pragmatic grounds, I think that choosing the macro- or the micro-explanation is not totally "a matter of taste". To be more charitable, it seems possible that what Sober really means by 'a matter of taste' is that scientific explanation depends on the question being asked and thus that there is no better explanation *per se*, rather than that each scientist has her personal taste of what it is a good explanation; if this is true, then my view has no conflict with his on this point. Therefore, he might agree with me that even in simple cases such as the peg-and-hole example in which an explanatory context is clearly delineated, there are good reasons to privilege one explanation over another, for example, to privilege the macro-explanation involving geometrical information over the fundamental one involving particle mechanics. Perhaps the contrast between my view and Sober's can be best recast by saying that Sober stresses the fact that there is no good explanation *per se*, while my view highlights the fact that in certain explanatory contexts, we may have good reasons to privilege one explanation over another.[27]

In sum, it is better to bear in mind that scientific explanation is, apart from an enterprise in search of truth, also a pragmatically driven activity in which the question being pursued, the context wherein the question is being raised and the cost-and-benefit considerations based on explanatory resources available concomitantly shape the space of potential answers that can be given. As such, the reductionist errs in claiming that better explanations always reside in the arena of lower-levels (e.g., the molecular level in biology) because these explanations are explanatorily deeper and more complete.

3.6 Conclusion

Not surprisingly, explanatory reduction, as one dimension of scientific practice, continues to play a significant role in science. Hence, in disagreeing with explanatory reductionists, I do not claim that explanatory reduction never happens in science but rather that it is not the *only* thing that happens in science. In this chapter, I evaluated the scope of explanatory reductionism by way of unearthing other dimensions of scientific practice (e.g., context considerations, extra information requirement and pragmatic factors that affect the evaluation and selection of explanations), in which a sweeping reductive strategy does not arise. The list of dimensions I presented is by no means exhaustive, but I think it is sufficient for my purpose.

Notes

1 According to Sarkar's terminology, ER is an epistemological thesis, which is distinguished from constitutive (ontological) and theory reductionism theses (Sarkar 1992).
2 Kaiser (2015) distinguished two subtypes of explanatory reductionism: (a) "a relation between a higher-level explanation and a lower-level explanation of the same phenomenon" (97) and (b) individual explanations—that is, given a relatively higher-level fact (or state, event, process, etc.), it can be reductively explained by a relatively lower-level feature (97). Because nowadays just a few authors hold the first subtype (e.g., Rosenberg 2006) while most embrace the second (e.g., Sarkar 1992, 1998, 2001, 2002, 2005; Hüttemann and Love 2011; also see Kauffman 1976; Wimsatt 1976, 2007), this chapter concentrates on the second subtype.
3 ER can be understood broadly or narrowly. Broadly construed, it is the claim that a higher-level feature can be better explained by a relatively lower-level feature (especially the feature at the molecular level; e.g., Rosenberg 2006); narrowly construed, it is the claim that features of Mendelian or classic genetics can be better explained by features of molecular biology (e.g., Waters 1990, 1994, 2000; Sarkar 1992, 1998). This chapter only concerns the broader construal.
4 Weber is an exception here, for his explanatory heteronomy thesis argues that the explanatory force of biology comes from natural laws of physics or chemistry. Thus, biology is explanatorily reduced to physicochemical laws (Weber 2005, 18–50).
5 This is contrasted with the reverse *many-to-one* multiple realization thesis, expressing that many different lower-level entities (or processes, pathways, mechanisms, etc.) can realize the same (or similar) higher-level phenomenon. This thesis has been discussed in Chapter 2, and we revisit it in this chapter.
6 Fang (2019a, 2020b) describes a modeling practice in which scientists attempt to tackle the context-sensitivity problem, namely, the *multilevel modeling practice* commonly employed in the biological, behavioral and social sciences. Importantly, this practice is nonreductive for typically numerous variables ranging over a few levels (e.g., molecular and cellular) are incorporated into a single model in order to explain a complex phenomenon of interest, and no level is privileged in the model.
7 In addition to the question of whether an appropriate context can be fully described molecularly, there is another question of whether such a description would have the same explanatory significance as the initial description—Section 3.5 addresses the latter question.
8 Alex Rosenberg (1997, 2006) expresses the same idea when appealing to "a total description of the fertilized egg" to reduce developmental processes to molecular processes. This chapter only considers Delehanty's argument.
9 Bechtel (2009) has a similar idea when he argues that, in constructing a good mechanistic explanation, researchers not only need to look down (i.e., decomposing a system into its components and interactions) but also need to look around (i.e., concerning how the system is organized) and look up (i.e., situating a mechanism in its context that might constitute a larger mechanism that regulates its behaviors). Bechtel's "looking up" corresponds to our top-down perspective in this chapter.
10 My discussion does not imply that scientists never explore the causal influence of certain contextual factors. They might do if, for example, they find that the phenomenon under consideration is so complicated that to fully understand it we might need to look out for somewhere else. Even so, this caveat is

still compatible with my position because by looking out for somewhere else, namely, by setting up a new inquiry, a new context is engendered again.

11 One may disagree with this verdict, arguing that the context simplifying strategy merely reflects our epistemic limits, rather than the ontological status of the world itself. Yet, as Laubichler and Wagner (2001) point out, it is unclear how one could uphold an ontological position that "cannot be transformed into actual scientific practice" (61). Given that the previous chapter has dealt with ontological reductionism and this chapter concentrates on epistemic (explanatory) reductionism, I say no more about the ontological issue in the following.

12 Note that the context independence argument concerns merely the degree to which the higher-level properties (patterns, mechanisms, phenomena, etc.) are *insensitive* to changes of their lower-level underpinnings. On the other hand, the higher-level properties in question (e.g., cellular) may also be contextually dependent on (or sensitive to) the still higher-level properties (e.g., organismal), a case compatible with the context-dependence argument proposed in this chapter.

13 Section 3.5 says more about *screening off*.

14 Note that in Huneman's (2010) original paper he treats flatness and sharpness as two different topological properties. However, as Arnaud Pocheville (personal communication) points out, they are, in fact, not different topological properties because there is a continuous transformation that can change one into the other; that is, they are the same topological property. Yet, it remains true that flatness and sharpness are two different *geometrical* characteristics that are relevant to how the two populations in our example behave differently.

15 This point is discussed in more detail in Chapter 7, where I discuss how a model can be explanatory.

16 One may worry that changing one population's shape feature can only be done indirectly, that is, by introducing mutation in the bacteria of that population. If this is true, then the intervention would exhibit the significance of the molecular level. However, I do not think this is necessarily the case. If, for example, we change the shape of the fitness landscape by changing the selective environment in which the bacteria reside, then the bacteria would remain unchanged. I thank Patrick McGivern for drawing my attention to this problem.

17 Remember what we have seen in Section 3.3, where the higher-level phenomena (e.g., the pattern formation regarding skeletogenesis) are partially independent of their lower-level molecular underpinnings.

18 The revised example also comes from Huneman (2010, 219–220), though with modifications.

19 Note that this does not necessarily mean that the underlying details are causally irrelevant. Rather, because of the partial independence of the geometrical properties from their underlying causal details with respect to explaining a phenomenon of interest, these underlying causal happenings are *screened off* by the geometrical properties. Section 3.5 says more about this point.

20 This point is highlighted by many authors in the context of non-causal explanation (e.g., Colyvan 2010; Baker and Colyvan 2011; Lyon 2012); Lange 2013; Pincock 2015; Baron, Colyvan and Ripley 2017 etc.).

21 One may argue that certain mechanisms can also be robust enough such that "they work always or for the most part in the same way under the same conditions" (Machamer, Darden, and Craver 2000, 3; also see Masel and Siegal 2009). Yes, that is true. Yet, it is also true that the degree of robustness

(or modal power) of geometrical properties is stronger than this because a geometrical property may be abstracted from a number of different systems that exhibit disparate underlying mechanisms; thus, changing one system's underlying mechanism does not necessarily result in changes in its geometrical properties—a point related to the partial independence of geometrical properties from their underpinnings discussed earlier. For a discussion of different degrees of modal force in the hierarchy of generalizations/laws (i.e., mathematical/logic generalizations/laws; meta-laws; first-order laws; subnomic facts, e.g., mechanisms; accidents), see Lange (2009).

22 This example is also drawn from Huneman (2010, 226–227), although with my reinterpretation and emphasis on its implications for reductionism.

23 I devoted one separate section to discussing context sensitivity simply because I think its importance deserves much more ink. Also, I think the argument based on context sensitivity is very powerful.

24 Unified explanation here means macro-explanations while disunified means micro-explanations (Sober 1999, 551).

25 Note that Rosenberg (2006) also holds an ideal image of explanatory reduction in which "there is a full and complete explanation of every biological fact, state, event, process, trend, or organization, and ... this explanation will cite only the interaction of macromolecules" (12).

26 Further argument for this claim can be found in Wimsatt's (1974, 1976, 2007 Chapters 4, 9, and 10) discussion of robustness, levels of organization, complex systems, and the like.

27 I thank Zhilin Zhang and Arnaud Pocheville for drawing my attention to this.

4 Are Laws the Only Thing That Matter?

4.1 Introduction

The last chapter has shown that we can find good explanations in biology even though they do not always reside at the supposed "ideal" level, that is, the molecular level. But there is another threat lurking on the horizon: admittedly, explanations in biology might not be reduced to some preferred level, but it might turn out that they are not genuine explanations at all, for they are either not laws or not law-based explanations. Unsurprisingly, known to philosophers of science for quite some time, the assumption underlying this threat is that only laws can provide genuine scientific explanations.

Undoubtedly, the notion of laws of nature is so deeply entrenched and universally accepted in science that any sensible philosopher of science cannot simply leave it aside. First, it concerns ontological issues like natural kinds, necessity, propensity, and capacity; second, it concerns epistemological issues like explanation, prediction, and inference; and, third, it concerns methodological issues like induction, deduction, and abduction. Our study of the natural sciences almost always starts by learning basic laws, like Newtonian gravitational laws, laws of thermodynamics, Mendelian genetic laws, and the economic law of demand and supply, to name just a few.

The problem surrounding laws of nature can be roughly divided into two relatively independent dimensions: (1) What are laws of nature? and (2) How can a generalization[1] (lawlike or non-law-like) be explanatory? The first concerns a metaphysical problem, wherein the notion of laws of nature is typically reductively defined in terms of some other notions such as *necessity*,[2] *capacities, dispositions* or *propensities*,[3] *members in deductive systems*,[4] and so on. The second concerns an epistemic problem, the problem associated with scientific explanation. In particular, the second problem concerns the pressing issue related to my book: "How can biological generalizations be explanatory given that they are not laws of nature?" These two problems are relatively independent because the question of why a generalization (lawlike or non-law-like) can be

DOI: 10.4324/9781003148029-4

explanatory can be properly answered without waiting for the availability of a clear-cut definition of laws of nature.[5] Note that the problem of "how can biological generalizations be explanatory given that they are not laws of nature?" can be further divided into two sub-questions: (a) the major question of this chapter: Are those generalizations that are explanatory in the biological sciences laws of nature? and (b) the key question for the following chapters: If, for example, they are not laws of nature, how can they be explanatory at all?

Given the relative independence of the epistemic problem from the metaphysical one and this book's primary theme (i.e., explanation), this chapter focuses on the epistemic dimension: How can a generalization (in the biological sciences) be explanatory even if it is not a law of nature? As said earlier, this question can be further divided into two sub-questions. So this chapter concentrates on the first sub-question: Are there laws in the biological sciences? For the second sub-question, a positive account (of how generalizations—in particular models—in the biological sciences can be explanatory) is not given until Chapter 7.

The layout of this chapter is as follows: Section 4.2 considers the far-reaching influence of the logical empiricist view on scientific explanation. Section 4.3 scrutinizes the long-standing debate over whether the biological sciences have laws of nature and argues that current attempts to establish distinctive laws in the biological sciences are unsuccessful. Upon denying the existence of laws in the biological sciences, Section 4.4 tentatively answers the question of why generalizations in the biological sciences that are nonlaws can be explanatory by suggesting that scientific models, rather than laws, do most (if not all) of the explanatory work. (The full answer is given in the following several chapters.)

4.2 The Far-Reaching Influence of the Logical Empiricist View

Philosophers' interest in laws of nature largely arises from the belief that to be explanatory a generalization must be a law of nature. This belief originated from the logical empiricist view (a view developed out of the Logical Positivist movement), which viewed laws as *universal* (spatially and temporally) generalizations that "are true not in virtue of pure logic or mathematics (are not analytic), but rather are true in virtue of the way the world is (are synthetic)" (Brandon 1997, S445). This *universal* character is guaranteed by the nomic necessity (as opposed to logical or mathematical necessity) property of laws, which distinguishes them from accidental facts that do not have such nomic necessity. Because of this property, another hallmark of laws is that laws can support counterfactuals while accidental facts cannot (Swoyer 1982; Fales 1990; Lange 2009).

Coupled with this view of laws of nature, the logical empiricists also had a philosophical account of scientific explanation, in particular, the deductive-nomological (DN) model developed by Hempel and

Oppenheim (1948). According to the DN model, "to explain is to provide a sound deductive argument for some explanandum, with at least one essential premise describing a natural law" (Weslake 2010, 274). Crucial to our current purpose is the implication that scientific explanations must invoke laws of nature and that the explanation of the phenomenon of interest is just the logical deduction of the phenomenon from the law(s) involved in the explanans plus certain initial and background conditions (i.e., antecedent conditions). Since this relationship is that of logical deduction, the relationship between the explanans and explanandum is also necessary, in the sense that the satisfaction of the explanans guarantees the truth of the conclusion (i.e. the explanandum).

On closer inspection, however, it is not difficult to find that the DN model of explanation is neither necessary nor sufficient for a legitimate account of scientific explanation (Scriven 1962). In particular, for an explanation to be genuinely explanatory, invoking laws is neither necessary nor sufficient. First, consider whether invoking laws is necessary for a generalization to be explanatory. Suppose there is such an explanation (E): "The low battery caused my computer to shut off". If asked why my computer turned off, (E) sounds perfectly good, for it gives us a legitimate answer that we are after. Also, it seems that this explanation does not involve laws or anything like that. So, one may conclude, it is unnecessary for an explanation to have laws to be explanatory. But proponents of the DN model would object to this, arguing that although this explanation appears not to involve any law, it in fact must *implicitly* involve laws so as to be complete and thus to be explanatory. For example, as the argument goes, it must implicitly involve the laws of electricity, or perhaps even laws regarding atoms that underlie the phenomenon of automatically turning off because of low battery. This argument, however, is not as convincing as it appears, for, as I argued in Chapter 3, explanation is inherently a pragmatic matter in which the question being raised, the context of raising the question and the cost-and-benefit considerations collectively determine what answers can best fit into the specific question raised. This means that the so-called explanation involving laws of electricity and atoms, although it may be true in essence, is simply not required for answering the question being raised (Wimsatt 1974).

Second, consider whether invoking laws is sufficient for a generalization to be explanatory. The flagpole problem is a good case in point with which we are all quite familiar:

> One can derive the length s of the shadow cast by a flagpole from the height h of the pole and the angle θ of the sun above the horizon and laws about the rectilinear propagation of light. This derivation meets the DN criteria and seems explanatory. On the other hand, a derivation ... of h from s and θ and the same laws also meets the DN criteria but does not seem explanatory. Examples like this suggest that at

least some explanations possess directional or asymmetric features to which the DN model is insensitive.

(Woodward 2014)

In short, both the derivation of s from h and θ and the derivation of h from s and θ invoke laws (i.e., geometrical laws and laws about the rectilinear propagation of light), yet only the derivation of s from h and θ seems explanatory. That is, even if a generalization invokes laws, it can still fail to be explanatory because it fails to capture the asymmetric feature of scientific explanation.[6] One diagnosis for why the generalization fails to capture the asymmetric feature is provided by Wesley Salmon (1989), who says that to explain a phenomenon one must cite the cause(s) of the phenomenon and that this is not reflected in the generalization (and not reflected in the DN model). Therefore, in our flagpole example, the reason a derivation of h from s and θ is not explanatory is that "the shadow does not cause the flagpole, and consequently cannot explain its height" (ibid., 47). In contrast, the reason a derivation of s from h and θ is explanatory is that "a flagpole of a certain height causes a shadow of a given length, and thereby explains the length of the shadow" (ibid.).[7]

Although more counterexamples can be found in the literature, cases like these discussed above are sufficient for our present purpose: for an explanation to be genuinely explanatory, invoking laws is neither necessary nor sufficient. Against this backdrop, many alternative approaches to scientific explanation have been proposed in the recent three decades or so that try to avoid, or at least loosen, the requirement for laws of nature. Among them are James Woodward's interventionist view, the stability view, the new mechanist view (sometimes also called the neo-mechanist view; for a comprehensive discussion of this view, see Craver and Tabery 2016), to name just a few. Woodward's interventionist view is that explanatory relations are those that, in principle, will support interventions and that will remain stable or invariant when we change various others things (e.g., background conditions), so whether a generalization can be used to explain has to do with whether it is *invariant* rather than with whether it is lawful (Woodward 1997, 2000, 2001, 2003, 2010).[8] The stability view is that the degree of stability of generalizations is a continuum, and therefore, some generalizations that are not as stable as strict laws can also be stable enough to explain phenomena (Skyrms 1980; Skyrms and Lambert 1995; Mitchell 1997, 2000; Lange 2000, 2004, 2009).[9] According to the new mechanist view, explanation is a matter of articulating the mechanisms that produce the phenomenon of interest, where mechanisms "are entities and activities organized such that they are productive of regular changes from start or set-up to finish or termination conditions" (Machamer, Darden and Craver 2000, 3; also see Glennan 2002, 2007, 2010; Machamer 2004; Bechtel and Abrahamsen 2005; Craver and Darden 2005; Craver 2007; Andersen 2012).

Nuance aside, all these views agree that explanation can be offered without necessarily invoking laws of nature, a case in which the explanatory work is typically done by generalizations that possess properties such as invariance, stability, and regularity (with respect to mechanisms).[10] And these properties in turn can be traced back to the type of phenomena under consideration. That is, although falling short of having the nomic necessity typically attached to laws of nature, generalizations that are invariant, stable, or regular enough are good candidates for explanation. This is by no means denying the importance or relevance of laws of nature in explaining phenomena; rather, it only claims that generalizations that are less stringent than laws can also properly assume explanatory labor. This is especially true in the special sciences in which laws are seldom found; in particular, it is true in the biological sciences where we can hardly find laws—a point I discuss in the next section.

Given this recognition that nonlaws can also do explanatory work, it might seem surprising that there are still a couple of major philosophers (e.g., Sober 1984, 1997, 2000; Weber 2005; Rosenberg 2006) who, explicitly or implicitly, insist on the view that to be explanatory is to be lawful. However, given the far-reaching influence of the logical empiricist view, it is not totally incomprehensible that its power is still lurking around. In particular, many philosophers nowadays still hold that laws must be in place in order to account for the explanatory success of the special sciences. For these philosophers, it seems we only have three options: (a) the biological sciences do have laws so they can be explanatory, (b) the biological sciences themselves do not have laws but they are based on certain underlying physicochemical laws (so their explanatory power comes from these laws), and (c) if the biological sciences neither have laws nor are based on physicochemical laws then philosophers owe us an account of why they can be so explanatory—since typically only laws can do explanatory work. The first option is held by Elliott Sober (1984, 1997, 2000, 2008, 2010) and Mehmet Elgin (2003, 2006),[11] the second by Alex Rosenberg (2001a, 2001b, 2006, 2012) and Marcel Weber (2005),[12] and the third by Rosenberg (2001c).[13] I disagree with all these options. In response to the first option, I argue in the next section that there are no laws in the biological sciences. In response to the third option, I claim in Section 4.4 that philosophically adequate accounts of explanation are available that do not invoke the notion of laws of nature. I do not discuss the second option since, on one hand, I show in what follows that there are no laws in the special sciences, and, on the other, I have already argued in Chapter 3 that scientific explanation is a pragmatic matter wherein lower-level explanation is not necessarily better than higher-level explanation (so there is no need to reduce).[14] In contrast with these authors, I tentatively suggest an account of scientific explanation which claims that scientific models, embedding the properties of invariance, stability or regularity, usually—if not all the time—do the explanatory work. (A full-fledged account of *why* and *how* they can do the explanatory job is developed in the following chapters.)

4.3 Are There Laws in Biology?[15]

Given the traditional view (i.e., the logical empiricist view) that to be explanatory is to be lawful, combined with the somewhat unpleasant fact that laws are nonetheless hard to come by in the biological sciences, a natural way to "save the phenomenon" that the biological sciences are explanatorily successful is to "redefine the meaning of scientific law" (Rosenberg 2001b, 743), namely, to cut the feet to fit the shoes. Among those who explicitly pursue this approach are Sober (1984, 1997, 2000, 2008, 2010) and Elgin (2003, 2006).

In philosophy of science, it is almost universally agreed that laws, at the very least, must be empirical, though it is also commonly recognized that this criterion alone cannot distinguish laws from nonlaws. However, Sober and Elgin disagree with this, arguing that the universally received empirical requirement is too strong to accept. By contrast, they suggest that we leave open the question of whether a law is an empirical or *a priori* truth (i.e., a mathematical truth). In particular, they contend that mathematical models in the biological sciences that can be true *a priori* are genuine laws of nature, such as Mendel's inheritance model (i.e., three laws of inheritance), Fisher's sex ratio model, Hardy-Weinberg equilibrium model, and so on. However, for several reasons, I think these models cannot count as genuine laws of nature.

4.3.1 *Confusion Between Mathematical Theorems and Mathematical Application*

I think there is confusion between mathematical theorems and mathematical theorems' applications when Sober and Elgin introduce some biological models as laws.[16] This can be shown with simple algebra. Elementary algebra tells us that if there are only two outcomes of an event E, and the probability of the occurrence A is p and the probability of the occurrence B is q, then for two independent and consecutive (or concurrent) events E_1 and E_2, the probability of composite outcome AA is p^2, BB is q^2, and AB is $2pq$. Now consider a biological model. The Hardy–Weinberg equilibrium model states that if four conditions obtain (viz., no immigration or emigration, random mating, no mutation, and no natural selection), and in generation one, the proportion of alleles P in the population is p, and the proportion of alleles Q is q, then the ratio of genotypes PP, QQ, PQ in the next generation will be p^2, q^2, and $2pq$. Comparing the basic theorem of elementary algebra described earlier with the Hardy–Weinberg equilibrium model, it is clear that the latter is just an application of the former (plus a certain number of initial conditions). Not surprisingly, the theorem of elementary algebra can also be perfectly applied to a plethora of other domains (e.g., economics, political science, anthropology, etc.) given certain initial conditions. For example, the "law of coin tosses" is another case in point. If two coins are tossed and each toss is independent of another and the probability distribution is $Pr(heads) = p$ and $Pr(tails) = q$, then the probability distribution of different outcomes is $Pr(2\ heads) = p^2$, $Pr(2\ tails) = q^2$, and $Pr(one\ head\ \&\ one\ tail) = 2pq$. With respect to this distinct application of the same mathematical theorem, can we therefore legitimately say that the "law of coin tosses" also constitutes a distinctive law of coins? If we insist that the Hardy–Weinberg equilibrium model is a distinctively biological law, nothing can prevent us from accepting that the "law of coin tosses" constitutes a distinctive law of coins as well. Furthermore, if we accept the "law of coin tosses" constitutes a distinctive law, we probably need to accept that the "law of pencil tosses" (with each side having the same probability of facing the ground) must constitute a distinctive law, and so on. The lesson is that, if we open the door to let mathematical applications to be genuine laws of nature, our cosmos would then have too many such cheap "laws". This is simply because mathematics can be applied almost everywhere.

4.3.2 Arms Race, the Evolutionary Contingency Thesis, and Ceteris Paribus Laws

The application of mathematical theorems includes mathematical models in biology such as the Hardy–Weinberg equilibrium model discussed earlier. As Nancy Cartwright points out, mathematical models are highly idealized entities such that they can only be true in idealized circumstances that are far away from reality. That is, they can only be *ceteris paribus laws* which are, strictly speaking, not true at all (Cartwright 1983). Ceteris paribus laws typically take the form that "other things being equal, then such and such". Nevertheless, when other things are not equal, which is more common than "other things being equal", these ceteris paribus laws simply break down.[17] This point can be best exemplified by the Mendelian inheritance model, including the law of segregation (the "first law"), the law of independent assortment (the "second law") and the law of dominance (the "third law"). These generalizations were once believed to be laws when first discovered by Mendel, but later, more and more exceptions were found (Rosenberg and McShea 2008, 46). In particular,

> once linkage was detected, it became clear that the second law is a rough and ready generalization with enormous numbers of exceptions, such as those arising from linkage. As for the law of segregation, geneticists now know cases in which segregation is unequal, in

> which one of the two alleles is preferentially transmitted to the next generation, the so-called segregate distorter alleles.
>
> (ibid., 46–47).

Therefore, Mendel's law of independent assortment can be at best understood as a "ceteris paribus law", stating that "if there were no linkage (and other things being equal), then separate genes for separate traits would pass independently of each other from the parents to the offspring" (ibid.).

Cases like Mendel's laws are not the exceptions; rather, they are very common in the biological sciences. When they are so common, there is a good story to tell why they are. According to the "arms race" metaphor[18] discussed by Rosenberg (2001c),

> [e]volutionary "design problems" have the reflexive character of ... dynamic strategic competitions in which every move generates a countermove so that conditions are never constant. From an early point in evolution, the environments that select for any one function began to change more rapidly. For each new solution to a design problem—each new functional trait—becomes part of the environment, setting a new design problem for other functional traits—within the same biological lineage or beyond it in a competing lineage. In the competition for limited resources endemic to the biosphere, any variation in the lineage of a gene, individual, group, or species which enhances fitness (that is, adaptation/function) will be selected for. Any response to such a variation within the heritable repertoire of the competitor gene, individual, group, or species will in turn be selected for by the spread of the first variation, and so on.
>
> (367)

This arms race implies that any ceteris paribus clause describing the condition under which a law applies is incomplete and can be filled in with further conditions. Moreover, the process of adding, or substituting, new conditions into/with the clause is hopelessly ceaseless. This sometimes means that "ceteris paribus laws" can only apply to highly restricted spatiotemporal regions that happen to fully satisfy the ceteris paribus clause.

The endless nature of the ceteris paribus clause can also be expressed from a different but related angle, an angle that highlights the contingent nature of "biological laws". In a similar vein, John Beatty (1995) puts forward his *evolutionary contingency thesis* (ECT), stating that "all generalizations about the living world: (a) are just mathematical, physical, or chemical generalizations (or deductive consequences of mathematical, physical, or chemical generalizations plus initial conditions), or (b) are distinctively biological, in which case they describe contingent outcomes of evolution" (218). The thesis is best formulated by DesAutels (2010):

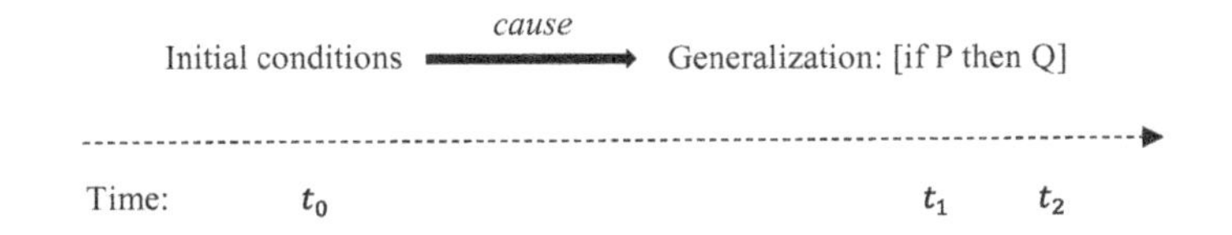

Figure 4.1 The evolutionary contingency thesis (ECT). This figure is modified from Sober (1997, S460).

"A set (I) of contingent initial conditions obtains at one time (t_0); this causes a generalization to hold true during some later temporal period (from t_1 to t_2)" (251; Figure 4.1).[19]

Beatty's key point is that the truth of (G) is guaranteed by the initial conditions (I), which themselves are just contingent outcomes of evolution that can be further changed during a slightly later temporal period from t_2 to t_3. And since evolution is an endless process as the arms race metaphor suggests, any generalization that holds true during a temporal interval is likely to break down later on. Hence there cannot be distinctive laws of biology; all we have are only spatiotemporally restricted contingent truths that appear and disappear at a certain time.

It is not difficult to identify the affinity between the ECT and the arms race metaphor introduced earlier. In a sense, Beatty's ECT can also be interpreted as the claim that the biological sciences only have "ceteris paribus laws", for holding true only when certain contingent initial conditions are present amounts to saying that "if such and such conditions were to be present in a certain spatiotemporal region (and other things being equal), then such and such". Though the arms race metaphor and the ECT emphasize different aspects (i.e., the endless and contingent nature of biological generalizations, respectively), the contingent nature of biological generalizations is grounded in their endless nature (which in turn is grounded in the endless nature of evolution); that is, if evolution were not ever-changing and ended at certain points, there might be noncontingent laws of biology.

Sober, however, disagrees with the ECT. He argues that there might be another generalization other than (G) that can be derived from Figure 4.2, which can be shown to be noncontingent. In particular, his noncontingent generalization takes the following form:

> (L) If (I) obtains at one time, then the generalization (G) that [if P then Q] will hold thereafter.
>
> (This sentence is adapted from Sober 1997, S460)

Sober claims that "the fact that the generalization [if P then Q] is contingent on I does not show that proposition (L) is contingent on anything" (ibid., S460). So (L) is a law.

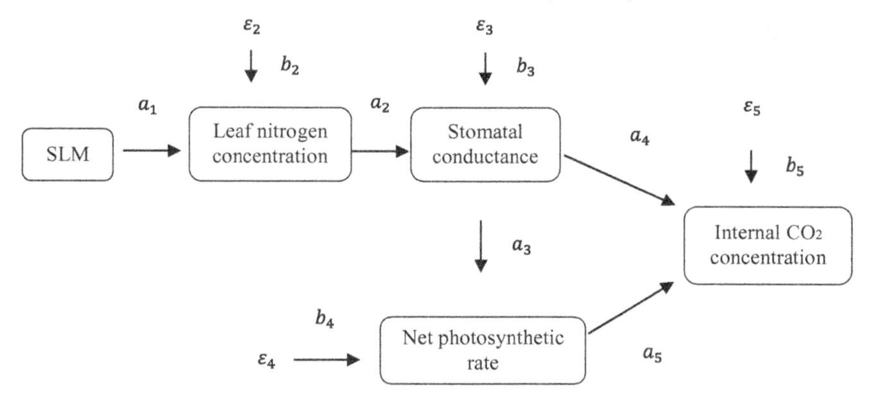

Figure 4.2 The proposed path model relating leaf morphology and leaf gas exchange. The letters with subscripts refer to the free parameters.

Sources: The figure is drawn from Shipley (2002, 131), with minor modifications. Figure used with permission.

However, I think Sober's response is problematic. To begin with, there is a problem associated with the controversial situation of ceteris paribus laws. The statement (L) strikes me as a ceteris paribus generalization, for it states precisely that if certain conditions are satisfied then something will happen. As said earlier, Beatty's ECT can be interpreted as the claim that the biological sciences only have ceteris paribus laws, which take the following form that "if such and such (initial) conditions were to be present at a certain spatiotemporal region (and other things being equal), then such and such". That is, Beatty's evolutionarily contingent generalizations correspond to ceteris paribus laws, which, for Beatty, are not genuine laws. Therefore, for Sober's argument, if we aim at establishing the fact that evolutionarily contingent generalizations such as (L) can count as laws, then we must also establish that ceteris paribus laws can count as genuine laws. However, as many philosophers have argued, ceteris paribus laws are not genuine laws,[20] and any attempt to hold them as genuine laws encounters a dilemma: ceteris paribus laws are either false or vacuous with respect to their empirical content.[21] This is because, on one hand, if we fill in the ceteris paribus clause with *complete* conditions that make the generalization true, then there will always be other interfering conditions that render the generalization false, and on the other hand, if we stipulate in the ceteris paribus clause that there are no other interfering conditions that render the generalization false, then the generalization simply becomes vacuously true (Reutlinger *et al.* 2015). Although there are attempts to avoid the dilemma (e.g., Fodor 1991; Hausman 1992; Pietroski and Rey 1995, etc.), it seems that they are not so successful (Earman and Roberts 1999; Woodward 2002; for a more comprehensive discussion of these attempts and their problems, see Reutlinger *et al.* 2015).

Although the issue surrounding ceteris paribus laws is still controversial and settling the debate (over whether they are genuine laws) is surely beyond this chapter's scope, it is sufficient here to point out that anyone holding ceteris paribus laws as genuine laws must face the substantial challenge (i.e., the dilemma mentioned earlier). So does Sober.

Second, Sober's response faces a serious problem: laws should not come so cheap. If (L) can be a law, then we can easily fabricate any number of *law-appearing* accidental facts by simply adding some initial conditions that render the accidental facts true. Suppose, for example, there is a strain of bacteria that live on the surface *S* during the time interval $t_i - t_j$. The corresponding generalization is that all individuals I_s of the strain live on the surface *S* during the time interval $t_i - t_j$. We may further find, suppose, that the relevant nutritional density *D* outside the *S* region is lower than the *S* region during the time interval $t_i - t_j$, and this may constitute a reason or initial condition for explaining why all I_s of the strain move to *S*. Finally, suppose that because of the strain of bacteria's consumption, after t_j the nutritional density *D* within and outside *S* will become the same and thus the strain of bacteria will move to somewhere else. Now we can construct a generalization like this:

> (L)* If (I) obtains at one time (i.e., if the nutritional density *D* in *S* is higher than its neighborhood), then the generalization (G) that [all individuals I_s of the strain live on the surface *S* during the time interval $t_i - t_j$] will hold thereafter.

In light of Sober's argument, we may say that (L)* constitutes a distinct law of nature. This conclusion, however, might taste unpalatable to many philosophers. Worse still, if we follow this line of constructing generalizations, we may find that countless alleged laws can be easily created. Furthermore, behind this toy example is the deeper worry that, given that many accidental facts have their own reasons for happening (i.e., initial conditions) that make them true during a certain time interval, it might turn out that accidental facts can constitute distinctive laws of nature.[22]

In sum, the arms race metaphor and Beatty's ECT make the claim that the biological sciences have their own distinctive laws dubious. Sober is suspicious of this conclusion, suggesting a different interpretation of the ECT, according to which the biological sciences do have their own distinctive laws. I have shown that Sober's treatment faces both a substantial challenge and a serious problem: (a) if evolutionarily contingent generalizations turn out to be ceteris paribus laws, then one must show how to avoid the dilemma surrounding holding ceteris paribus laws as genuine laws, and (b) it may lead to the upshot that laws are so easy to come by so that many accidental facts can count as laws.

4.3.3 Why Worry About Laws in the Biological Sciences?

The foregoing two subsections have shown that the attempts to establish distinctive laws in the biological sciences are not successful. Some attempts are hindered by confusion between mathematical theorems and their applications, and some are undermined by serious problems (e.g., blurring the contrast between laws and accidents). Perhaps it is time to consider the motivation behind those who worry too much about laws in the biological sciences. There are good grounds to ask why we should care about the laws of nature in the biological sciences *at all*. As discussed in Section 4.2, the obsession with laws is partially driven by the nomological commitment stemming from the Logic Empiricist view, which holds that only those explanations that explain phenomena by means of deducing them from natural laws can constitute genuine scientific explanations. Although nowadays few hold the view that to explain a phenomenon is to deduce it from laws, the assumption that explanation must appeal to laws remains an important tenet for some philosophers (e.g., Rosenberg 2001c, 2006; Weber 2005; etc.).

Against this background, it is understandable that there are still many philosophers worrying too much about laws in the biological sciences. On the one hand, philosophers find that there are no laws in the biological sciences resembling those found in physics, those laws that are believed to be universally true and exceptionless. These alleged universally true and exceptionless laws are believed to play a central role in explanation in physics. On the other hand, it is a plain fact that the biological sciences are very successful in explanation (and prediction). So there seems to be a tension between the absence of laws and explanatory success. Hence, here is a need to account for why the biological sciences (or, more generally, the special sciences) are so successful in explaining phenomena given that they do not have laws resembling those in physics. As mentioned in Section 4.2, to deal with the tension and to account for the explanatory success, it seems we only have three options: (a) the biological sciences do have laws so they can be explanatory, (b) they do not have laws and their explanatory power is based on physicochemical laws, and (c) they do not have laws but then we need an account of why they can be so explanatory.

Unsurprisingly, many philosophers take the first option. In doing so, the motivation is quite clear: we must appeal to *something* to account for the explanatory success of the biological sciences, and laws seem to be the most promising candidates in our checklist. Even for those taking the second or the third option, the motivation is almost the same as those taking the first. For example, when Weber (2005, 18–50) advances the explanatory heteronomy thesis (according to which the explanatory force of the biological sciences solely comes from applying physicochemical laws), he is clearly motivated by the assumption that only laws can explain. Also, the motivation is explicitly expressed by another major player in this field, namely, Rosenberg (2001c, 370), who has been worrying about laws in the biological sciences for decades; he states that "denying that there are laws in biology is unattractive because it comes with a concomitant obligation to provide a new account of how explanation proceeds in biology".

By and large, I think Rosenberg's worry is worth ruminating on, for we do need an account of how explanation really works in biology. In fact, I take Rosenberg's worry very seriously, for the next several chapters of this book are devoted to accounting for how explanation proceeds in the biological sciences. However, I think his judgment that denying laws in biology is unattractive is itself unattractive; on the contrary, I suggest that denying laws in the biological sciences is a very attractive approach along which new paths of exploration will be opened up, a point I discuss in detail in the next section.

4.4 An Alternative to Laws: Models

As mentioned earlier, Rosenberg claims that we have to provide a new account of how explanation proceeds in the biological sciences if they do not have laws. I think that, although accounts of how explanation proceeds in the biological sciences are needed, we should not worry too much. This is because many tenable accounts that step far away from the logical empiricist commitment (i.e., explanation must appeal to laws) are available now, including Woodward's interventionist account, the stability view, and the new mechanist view, briefly discussed in Section 4.2. Although these views differ from one another in many aspects, they, by and large, all agree that explanation can be offered without necessarily invoking laws of nature. In short, they all detach themselves from the lawfulness requirement, sharing the view that nonlaws can also do explanatory work.

All these views emphasize the properties of phenomena under consideration (such as invariance, stability, regularity, etc.) that can be employed for explanation (so they are emphasizing the *ontic* side of explanation), but they do not highlight sufficiently the *common and typical form of representation* of these phenomena (so they do not stress the *epistemic* side of explanation sufficiently). To fill in this gap, I claim that the most typical forms of representation in the biological sciences are models. In other words, we typically use models in the biological sciences to represent the target systems in order to explain, predict, control, manipulate, and so on (so my approach to explanation is both ontic-oriented and epistemic).[23] Proponents of the new mechanist view mention this idea, for example, they say that "explanations in the life sciences frequently involve presenting a model of the mechanism taken to be responsible for a given phenomenon, and such explanations depart in numerous ways from nomological explanations commonly presented in philosophy of science" (Bechtel and Abrahamsen 2005, 421). Also, talk of models is found throughout Woodward's seminal book (2003), although he is dealing with scientific explanation in general rather than model explanation in particular.[24] Although a number of views on how models can be explanatory have been developed in the literature,[25] a substantive account on both *how models can be explanatory* (*how-question*) and, more importantly, *in terms of what models can explain in this or that particular way* (*what-question*) remains to be seen. Answering the *what-question* requires an articulation of the model–world relationship, which I discuss at length in Chapters 5 and 6.

Note that my approach is built on the previous views (i.e., the interventionist view, the new mechanist view, and the stability view) rather than conflicting with any of them. First, as will become clear in the following chapters, my approach also regards models as having the property of invariance, which entails that they can be employed to answer Woodward's *what-if-things-had-been-different* questions. Second, some models represent mechanisms of the biological phenomena in question, so these models can be classified into the category of mechanistic models.[26] Third, depending on different kinds of models and different goals of modeling, the degree of stability (i.e., invariance) of models can vary from case to case, so some models may be more stable than others with respect to certain specific aspects.

If we agree that the common and typical form of representation in the biological sciences is the model, then the central issue comes down to such a *general* question: *how* models can be explanatory. This central issue is left for the following chapters; for the moment, let us first consider a factual issue: Do many explanations in the biological sciences take the form of models? I think the answer is in the affirmative, and the shift of focus in the literature on scientific explanation toward model-based approaches supports this answer (Batterman 2002b; Woodward 2003; Giere 2004; Craver 2007; Godfrey-Smith 2006; Bokulich 2008, 2011, 2012; Strevens 2008; King 2016). Suffice here to simply mention a handful of examples that played, and continue to play, pivotal roles in the practice of the biological sciences: Mendel's inheritance model, Fisher's sex ratio model, Hardy–Weinberg equilibrium model, Lotka–Volterra model of predation, the "lock-and-key" model and the induced-fit model of enzymes, the double helix model of DNA, the evolutionary game model or theory (e.g., Maynard Smith and Price's Hawk–Dove game (Smith and Price 1973), the island model of population differentiation, and so on and so forth).

To set the stage for the following chapters, I outline two models as examples in the following subsections: one is a mathematical model (also a biological model)—the leaf gas-exchange model—and one is a concrete model (also a nonbiological model)—the San Francisco Bay model. There are reasons why I choose these two models. To begin with, they are neither too complicated nor too simple. Second, they exclusively represent two kinds of models in terms of Michael Weisberg's distinction: mathematical models and concrete models (Weisberg 2013).[27] Third, the leaf gas-exchange model is a good exemplar in the discourse of causal modeling[28] that is relevant to developing my views in the following chapters, while the San Francisco Bay model, as a nonbiological model, provides a good place to examine how far my views built on biological models can go. Given that the Bay model has been investigated by Michael Weisberg (2013) in favor of his similarity view of models, a reexamination of the model in this book has another advantage: it shows that by looking closely at the model a different view of models can be developed.

4.4.1 A Mathematical Model: The Leaf Gas-Exchange Model

To understand the causal mechanism underlying leaf gas-exchange, Shipley and Lechowicz (2000) proposed a path model involving five variables: specific leaf mass (SLM: leaf dry mass divided by leaf area, g/m^2), leaf organic nitrogen concentration (mmol/m^2), stomatal conductance to water (mmol/m^2 per s), net photosynthetic rate (μmol/m^2 per s) and internal CO_2 concentration (μl/l). This model is expressed in Figure 4.2.

In Figure 4.2, directed lines represent causal relations, and square boxes are causal relata. a_x and b_y are free parameters, and ε_n denotes an error variable representing other unmodeled causes of the variable into which it points.

As the figure shows, this model represents a causal structure among various variables that are directly or indirectly connected. The constructed model at this stage, however, constitutes at best a hypothesis such that modelers need to check whether it appropriately captures the phenomenon of interest. That is, modelers need to test its *goodness of fit* to the reality (or the target system). The testing regarding this model usually involves a mathematical method called the *maximum likelihood estimation method*, which is discussed in the next chapter.

Upon carefully scrutinizing this model and its testing method, the readers will find that the model fits its target system in a holistic manner, wherein one organized whole fits another organized whole; this point is developed in Chapters 5 and 6. Also, once the model has been confirmed to be good enough (i.e., holistically fits its target system very well), we are able to answer why the model is explanatory: because the model shows how changes in one variable can be systematically associated with changes in other variables. The details of this point are developed in Chapter 7.

4.4.2 A Concrete Model: The San Francisco Bay Model

In 1950s, many people started to worry about the fragility of the water supply in the San Francisco Bay area. To solve this problem, an ambitious plan was proposed, that is, to dam up the Bay, which "would not only supply San Francisco with nearly unlimited drinking water, but also revolutionize the area's transportation, industrial, military, and recreation infrastructure" (Weisberg 2013, 1). However, objectors worried that doing so would cause serious problems; for example, it would "destroy commercial fisheries, render the South Bay a brackish cesspool, and create problems for the ports of Oakland, Stockton, and Sacramento" (Jackson and Peterson 1977; cf. Weisberg 2013, 1).

To settle this dispute, the Army Corps of Engineers was charged with investigating the influence of the proposed plan by building a massive hydraulic scale model of the Bay system (Weisberg 2013, 1–2). The model description is here:

> Constructed to a horizontal scale of 1:1,000 and a vertical scale of 1:100, the original model included all of San Francisco Bay proper, San Pablo Bay, Suisun Bay to the confluence of the Sacramento and San Joaquin Rivers, and 17 miles of the Pacific Ocean beyond the Golden Gate.... The model presently occupies an area of about one acre, being some 340 feet long in a north-south direction and measuring about 450 feet east to west.... The model was constructed of precast, lightweight, reinforced concrete slabs, generally 12 feet by 12 feet in plan, supported at each corner by levelling screws.... The geometric scales fixed the following scale relationships: slope, 10:1; velocity, 1:10; time, 1:100; discharge, 1:1,000,000; and volume, 1:100,000,000. The salinity scale ratio required for an investigation of this type is 1:1.
>
> (Huggins and Schultz 1973, 12)

Once the model was built, it was adjusted to accurately reproduce several measurements of the parameters such as tide, salinity, and velocities actually recorded in the Bay. This involved adjustment of the tidal apparatus to reproduce the ocean tide of 21 September 1956, to obtain proper reproduction of tidal phenomena in the lower portion of the Bay system, followed by adjustment of the frictional resistance of the model bed until tidal elevations and times throughout the model were in agreement with prototype mean tide data (Huggins and Schultz 1967, 11) and so on. A grand total of about 25,000 copper stripes used to reproduce roughness were embedded in the model during and after construction (ibid., 11). The Bay system was so complex that it took 14 to 15 months in total to adjust the roughness in order to "obtain a reasonably fair reproduction of the half-hourly velocities and hourly salinities which were measured in the prototype in the deep-water channels" and to reproduce "accurate flow distributions in various cross sections of the model" (ibid., 11).

Fortunately, the Bay model worked very well after adjustment:

> Agreement between model and prototype for the verification survey of 21–22 September 1956, and for other field surveys, was excellent.

> Tidal elevations, ranges and phases observed in the prototype were accurately reproduced in the model. Good reproduction of current velocities in the vertical, as well as in the cross section, was obtained at each of the 11 control stations in deep water and at 85 supplementary stations. The salinity verification tests for the verification survey demonstrated that for a fresh-water inflow into the Bay system ..., fluctuation of salinity with tidal action at the control points in the model was in agreement with the prototype.
>
> (Huggins and Schultz 1967, 11)

After the adjustment, it was time to evaluate the proposed plan. The plan involved "a 600 foot wide, 4 mile long earth and rock dam that stretched from San Quentin to Richmond, and a second barrier 2000 feet wide and 4 miles long, just south of the Bay Bridge, connecting San Francisco to Oakland" (Weisberg 2013, 9). The Army Corps investigated this plan by building scaled barriers, "adding these to the Bay model, and measuring the changes in current, salinity and tidal cycles" (ibid., 9). The investigation showed that it would considerably reduce water-surface areas (62 percent in South Bay and 75 percent in San Pablo Bay), reduce the velocities of currents (from 20–30 to 0.2–0.3 ft/sec) in most of South San Francisco Bay, reduce the tidal discharge through the Golden Gate during the tidal cycle, worsen the poor dispersion and flushing characteristics in the Bay area, and so on (Huggins and Schultz 1973, 19). Given these disastrous consequences, the Army Corps had good reasons to reject the proposed plan (Weisberg 2013, 9).

This is the outline of the Bay model. As shown in Chapter 5, regarding the leaf gas-exchange model, and in Chapter 6, when looking closely at the Bay model and its testing procedure, the Bay model fits its target system in a holistic manner. Or generally speaking, the model–world relationship is a holistic matter.

4.5 Conclusion

This chapter claimed that to be explanatory is not necessarily to be lawful, because things other than laws of nature can also properly do the explanatory work; these explanatory tools are scientific models that are widely used in the biological sciences. To better dispel laws and thus establish models, Section 4.2 considered the far-reaching influence of the logical empiricist commitment regarding scientific explanation, in which the DN model of scientific explanation is always assumed. The influence of the logical empiricist commitment, and hence the obsession with laws, is clearly reflected in the apparent tension in the biological sciences: on the one hand, the biological sciences do not have typical laws, and on the other, they explain phenomena of interest very well. One way to release this tension is to stretch the meaning of laws to such an extent that it fits the logical empiricist commitment that the biological sciences proceed very well because they have a special form of laws. However, Section 4.3 argued that this approach is unsuccessful because (a) it simply confuses mathematical theorems with their applications and (b) it faces both a substantial challenge (i.e., how to deal with the falsity/vacuousness dilemma if evolutionarily contingent generalizations are ceteris paribus laws) and a serious problem (i.e., blurring the boundary between laws and accidental facts).

The denial of laws of nature leaves us with a conundrum: How can the biological sciences explain if they lack the commonly assumed vehicle of explanation, namely, laws of nature? Many plausible approaches have been developed that step far away from the logical empiricist commitments, which share the basic idea that generalizations that are invariant, stable or regular enough can do explanatory work. Based on these alternative approaches, I found the need to develop a new account that both inherits the merits of these alternatives and fills in the lacunae neglected by them. In particular, I suggested that scientific models, as the common and typical form of representing phenomena, typically do the explanatory work in the biological sciences. A full-fledged account of how biological models can be explanatory is given in Chapter 7.

Notes

1 *Generalizations* should be understood broadly throughout this book to include not only linguistic statements but also models (concrete or mathematical) that function as forms of expressing knowledge.
2 See Armstrong (1978, 1983, 1991, 1993), Dretske (1977), Tooley (1977, 1987), Bird (2001, 2002, 2004), and Lange (2009), among others.
3 See Swoyer (1982), Cartwright (1983, 1989, 1999), Ellis and Lierse (1994), Ellis (2001, 2002, 2010), and Bird (2005a, 2005b), among others.
4 See Ramsey (1978/1928), Lewis (1973, 1983, 1986, 1994), Earman (1984), and Loewer (1996), among others.
5 Admittedly, for those who believe that only laws of nature can be explanatory, the epistemic problem is not independent of the metaphysical one. However, this book aims to establish the independence by showing that (a) there are no laws in the biological sciences (this chapter) and (b) models rather than laws do most (if not all) of the explanatory work in the biological sciences (Chapter 7).
6 However, van Fraassen questions the supposed asymmetric feature of explanation by arguing that there might be cases in which the length of the shadow plus the angle of the sun can be used to explain the height of the flagpole (1980). However, even if that is true in some cases, it does not impact our conclusion that the DN model cannot shed light on the asymmetric feature of explanation in some other cases. Since this chapter is not about scientific explanation *per se*, I do not discuss this problem in the following. I thank Patrick McGivern for letting me notice van Fraassen's point of view.
7 For a more comprehensive discussion of the problems the DN model faces, see Woodward (2014).
8 Chapter 7 says more about Woodward's approach.

9 Note that Skyrms (1980; also Skyrms and Lambert 1995) uses *resiliency* rather than *stability* to refer to the property of a law or generalization (for a discussion see Woodward 2003, 299–302).

10 Mechanisms have the property of *regularity* in the sense that they are regular; that is, "they work always or for the most part in the same way under the same conditions" (Machamer, Darden and Craver 2000, 3).

11 Also see Cooper (1996), Brandon (2006), Lorenzano (2006), Press (2009), Morgan (2010), Dorato (2012), Sachse (2012), Haufe (2013), Gouvêa (2015), and others who hold that the biological sciences do have laws.

12 Rosenberg's attitude toward this problem is a bit complicated, for he denies there are laws in the biological sciences except the principle of natural selection; the principle is the only law, which is built purely on a physicochemical basis (for a detailed discussion see Rosenberg 2006, 177–200). For simplicity, throughout this chapter, I assume that Rosenberg holds that there are no laws in the biological sciences. Weber (2005, 18–50) upholds an explanatory heteronomy thesis in biology according to which there are no laws in the biological sciences and the explanatory force of the biological sciences solely comes from applying physicochemical laws to the biological domains.

13 See Note 12.

14 Note that my position can be compatible with the assumptions (i.e., the biological sciences do not have laws and biological processes are based on their underlying physicochemical processes/laws) but not with the implication (i.e., the explanatory power of the biological sciences comes from these physicochemical processes/laws) of the second option. This is because even if biological processes are based on physicochemical laws, whether we should always appeal to these laws to explain biological phenomena is another question. According to my view developed in Chapter 3, we sometimes privilege higher-level over lower-level explanations, so it follows that the explanatory power of the biological sciences does not always come from these underlying physicochemical laws (for a discussion about reducing biological explanation to physicochemical explanation, see Sarkar 1998).

15 Note that there is a parallel debate over whether there are distinctively ecological laws, in which the proposed candidates for laws are the Malthusian law of exponential population growth, the allometries of macroecology, the rules of stoichiometry, and so on (see Murray 1992, 1999, 2000; Quenette and Gerard 1993; Lawton 1999; Turchin 2001; Berryman 2003; Colyvan and Ginzburg 2003; Ginzburg and Colyvan 2004; Lange 2005; O'Hara 2005; Dhar and Giuliani 2010; Raerinne 2011; Sagoff 2016; etc.). Since what will be said in this chapter about biological laws in general is also true about ecological laws in particular, this chapter focuses on Sober and Elgin's arguments.

16 Several other authors also raise issues with Sober and Elgin's *a priori* laws, for example, Rosenberg (2001b), Press (2009), and Raerinne (2011), among others.

17 Some authors argue that, apart from ceteris paribus laws, we also have non-ceteris-paribus laws, such as certain symmetry principles in physics (e.g., the conservation laws; see Earman 2004; Lange 2007, 2009).

18 This metaphor originally comes from Richard Dawkins (1986).

19 Beatty's idea is influenced by Stephen Jay Gould who expressed the idea in the 1980s (see Gould 1989).

20 See Cartwright (1983); Schiffer (1991); Earman and Roberts (1999); Earman, Roberts and Smith (2002); and Woodward (2002) for relevant discussion.

21 See Lange (1993), Earman and Roberts (1999) and Earman, Roberts and Smith (2002) for relevant discussions.

22 DesAutels (2010) recently made a similar argument against Sober and Elgin as I present here. His idea is that if we let some evolutionarily contingent generalizations be laws, then any biological phenomenon can count as a law and the interesting contrast between laws and accidental facts collapses.

23 Note that there are cases where the model itself is the target system (e.g., model organisms), and there are cases where there is no clear target system to represent at all (e.g., cellular automata or minimal models). For a discussion of targetless models, see Weisberg (2013, 130–132), and for a discussion of minimal models that are not representative see Batterman (2002a, 2002b, 2009, 2010), and Batterman and Rice (2014).

24 Note that Woodward uses the term *models* to refer to two different categories of models: philosophical models (e.g., the DN model of scientific explanation) and scientific models (e.g., causal models in economics).

25 For example, see Batterman (2002a, 2002b, 2009), Bokulich (2008, 2011, 2012), Kennedy (2012), Rice (2012, 2015), Batterman and Rice (2014), and Rohwer and Rice (2016), among others.

26 For a discussion of mechanistic models and mechanistic modeling, see Weisberg (2013, 150–152).

27 Note that in Weisberg's framework mathematical models have a subcategory: computational models. Ontologically speaking, however, computational models are just mathematical models, although they function differently in practice (Weisberg 2013, 20–23). For simplicity, in this book, computational models are not discussed.

28 For a comprehensive discussion of this model and its testing methods, see Shipley (2002).

5 Models That Matter

The Similarity View

5.1 Introduction

The last chapter claimed that the biological sciences can be explanatory even without laws, because the alternative to laws, that is, models, representing the property of invariance (or stability, regularity) usually do the explanatory work. Although it is an unadorned fact that the biological sciences are abundant in models, there is still a question of how those models can be explanatory. This question can somehow be reframed as "What is the model–world relationship whereby we can use the model to explain the world?" Namely, the question has two facets, one concerning the model–world relationship and the other concerning how a human-made model explains a target phenomenon in the world. Importantly, the two facets are connected in a way that getting a handle on the first facet helps us make sense of the second. Yet the question has become very pressing recently given that many conflicting accounts are available in the literature, each of which emphasizes some aspects of models and modeling.

In particular, the similarity account of models, first developed by Ronald Giere and then further elaborated by Peter Godfrey-Smith and Michael Weisberg, has recently gathered much force in the literature. This account started with the intuition that a model should at least be similar to its target system so as to be explanatory, matured with the idea that a model should be similar to its target *in certain respects* and *to certain degrees* and culminated in the sophisticated weighted feature-matching account developed by Weisberg. The similarity relationship is also termed the representational relationship, whereby the model represents the target system. Although this account is largely right in capturing the intuition that a model should be at least similar to its target system, it errs in capturing the intuition in the wrong direction: it describes the model (and its target system) as a set in which elements (i.e., features) are independent of one another.

DOI: 10.4324/9781003148029-5

5.2 Giere and Godfrey-Smith's Similarity Account

We may ask why a model can be explanatory given that it is only a "miniature" of the reality that it intends to mimic. Not surprisingly, due to the complexity, inaccessibility, uncomputability, or sometimes even capriciousness of the target system, a model can hardly be an all-encompassing description (or representation) of the reality or even a faithful description. Moreover, the situation can become even more vexed if we are told that, as William Wimsatt says, most scientifically useful models are simply false (due to idealizations, simplifications, distortions, false assumptions, etc.). At any rate, it sounds self-contradictory to say that false models can act as means to truer theories (Wimsatt 2007). How can that be possible?

A natural answer to this question is that a model is explanatory because it bears a similarity relationship *in certain respects and to certain degrees* to its target system. So knowing the relevant respects and degrees of the model may help us know the corresponding respects and degrees of the target. This is just the original idea of Giere; as he says,

> [t]he appropriate relationship, I suggest, is *similarity*. Hypotheses, then, claim a *similarity* between models and real systems. But since anything is similar to anything else in some respects and to some degree, claims of similarity are vacuous without at least an implicit specification of relevant *respects and degrees*. The general form of a theoretical hypothesis is thus: Such-and-such identifiable real system is similar to a designated model in indicated respects and degrees.
>
> (Giere 1988, 81; author's emphasis)

Godfrey-Smith (2006) expresses more or less the same idea:

> Many of the special features of model-based science come from the role played by resemblance relations between model system and target. Philosophers tend to distrust resemblance relations because they are seen as vague, context-sensitive, and slippery. Here, those features of resemblance are indeed important, but they are not necessarily problematic. They are the source of both distinctive strengths and weaknesses of model-based science.
>
> (733)

However, because neither Giere nor Godfrey-Smith specifies clearly what respects should be counted and what degrees should be tolerated when building and testing models, some find the early version untenable (e.g., Suárez 2003). To deal with this problem and fill in the lacuna, Giere (2004, 2010) later developed a more sophisticated similarity account, according to which scientific representation is a four-place activity:

> Shifting the focus to scientific practice suggests that we should begin with the activity of *representing*, which, if thought of as a relationship at all, should have several more places. One place, of course, goes to the agents, the scientists who do the representing. Since scientists are *intentional* agents with goals and purposes, I propose explicitly to provide a space for purposes in my understanding of representational practices in science. So we are looking at a relationship with roughly the following form: *S* uses *X* to represent *W* for purposes *P*.
>
> (2004, 743; author's emphasis)

Hence, apart from models and the target systems themselves, the activity of scientific representation (and scientific explanation) should also incorporate factors such as the role played by scientists and the intentions those scientists have regarding modeling. Incorporating these factors is important partly because similarity is a symmetric relationship in which if *A* is similar to *B* then *B* is also similar to *A*, whereas scientific representation (and explanation) is an asymmetric process wherein one can only claim that the model represents the world and thus is similar to the world in certain aspects but not vice versa. When including intentions, the modeling process is naturally asymmetric as the scientists only intend to use the model to represent the target.

In short, the early version of the similarity view says that for a model to be explanatory, it not only needs to be similar to its target in certain respects and to certain degrees but also considers the role played by modelers and their intentions or goals. On the basis of this early version, a more sophisticated version has been proposed very recently.

5.3 Weisberg's Weighted Feature-Matching Account

Weisberg (2013) has recently proposed an account that is "intended to be sensitive to how scientists represent the world with models and to how their representational goals and ideals shape the standards of fidelity that they apply to models" (135). In particular, it aims to answer questions about "what similarity supervenes on, how it depends on context, how similarity judgments are to be evaluated" (ibid., 143). Elsewhere, he says his account could let us "capture the similarity judgments made by scientists" (ibid., 155). In a word, it attempts to not just generally spell out what the notion of similarity means but, more important, also aims to capture similarity judgments made by scientists[1] given how they choose a feature set, a weighting function, and assign values to weighting parameters.[2]

Weisberg (2012, 2013) calls his approach to similarity the *weighted feature-matching* account. He borrows the basic ideas from psychologist Amos Tversky's *contrast* account of similarity, which claims that the similarity of objects *a* and *b* depends on the features they share and the features that they do not share. In light of this, Weisberg proposes his own account of similarity:

$$s(m,t)=\frac{\theta f(M_\alpha \cap T_\alpha)+\rho f(M_m \cap T_m)}{\theta f(M_\alpha \cap T_\alpha)+\rho f(M_m \cap T_m)+\alpha f(M_\alpha - T_\alpha)+\beta f(M_m - T_m)+\gamma f(T_\alpha - M_\alpha)+\delta f(T_m - M_m)}, \quad (5.1)$$

where $f(x)$ refers to the weighting function; α, β, γ, δ, θ, and ρ denote weighting terms (parameters); subscripts a and m stand for attributes and mechanisms, respectively; M denotes the model; and T, the target. $(M_a \cap T_a)$ stands for attributes shared by the model and the target, $(M_a - T_a)$ represents attributes that the model has that the target does not have and $(T_a - M_a)$ attributes that the target has that the model does not have. The same story goes for mechanisms m. Attributes and mechanisms as a whole are called features of the model and the target.

Then an interpretation for this equation is given. First, there must be a feature set Δ, and the set of features of the model and the set of features of the target are defined as sets of features in Δ. Primarily illuminated by Giere's *four-place activity* account, Weisberg claims that the elements of Δ are determined by "a combination of context, conceptualization of the target, and the theoretical goals of the scientist" (ibid., 149); that is, "there is no context-free answer to this question, but part of the answer lies in the modeler's intended scope. The modeler's intended scope takes into account the research question of interest, the context of research, and the community's prior practice" (ibid., 149). What is more, the contents of Δ might change through time as science develops, which, in turn, might result in a reevaluation of the established model–world relationship (ibid., 149).

Second, consider the values of the weighting parameters α, β, γ, δ, θ, and ρ. On Weisberg's account, different kinds of modeling require different weighting parameters. For example, if what interests us is the *minimal modeling* which concerns merely the mechanism responsible for bringing about the phenomenon of interest, the goal of this modeling is written as[3]

$$\frac{|M_m \cap T_m|}{|M_m \cap T_m| + |M_a - T_a| + |M_m - T_m|} \to 1, \quad (5.2)$$

where $|M_m \cap T_m|$ has a high value while $|M_a - T_a|$ and $|M_m - T_m|$ have low values (ibid., 151).

Finally consider the weighting function $f(x)$, which tells us the relative importance of each feature in the set Δ. In Weisberg's view, scientists in most cases have in their mind some subset of the features in Δ, which they regard as especially important. Hence, some features are weighted more heavily, and others would simply be equally weighted. According to Weisberg, the background theory determines which features in Δ should be weighted more heavily. In cases in which the background theory is not rich enough to make these determinations, deciding which features should be weighted more heavily is, in part, an empirical problem.

This is Weisberg's account in an abstract form. To give you a taste of how Weisberg's account works and to pave the way for showing how it differs from the actual modeling practice in the next section, let me consider a toy example. Imagine in the real world there is a causal process between the number of functional copies of a gene (G) and the rate (E) at which some enzyme is produced.[4] Based on empirical data, scientists build a model (Figure 5.1).[5]

After building this model, scientists attempt to test whether the model captures the target—or, to use Weisberg's terminology, whether the model is maximally similar to the target. Suppose for the moment that Weisberg's account can be applied to this simple case.[6] This modeling falls into the category of minimal modeling,[7] in which scientists are interested in the mechanism responsible for bringing about the phenomenon of interest. Therefore, we can apply Equation 5.2 to this case, in which the only mechanism involved is a causal process between G and E. Adding back those simplifications about weighting functions and weighting terms to Equation 5.2, we obtain Equation 5.3:

$$S(m,t) = \frac{\rho f(M_m \cap T_m)}{\rho f(M_m \cap T_m) + \alpha f(M_a - T_a) + \beta f(M_m - T_m)}. \tag{5.3}$$

For simplicity, we stipulate $\rho = \alpha = \beta = 1$. Since minimal modeling concerns only the mechanism responsible for bringing about the phenomenon, we weight $(M_m \cap T_m)$ very heavily and $(M_a - T_a)$ and $(M_m - T_m)$ lightly; for example, suppose the values of weighting functions for them are 0.81, 0.05 and 0.09, respectively. Thus, we finally achieve a similarity value: $0.81/(0.81 + 0.05 + 0.09) = 0.85$, which is very near the maximal similarity value 1.00. Therefore, if the testing process is correct, scientists then can conclude that their model is good enough.

Figure 5.1 A biological model of a causal process.

5.4 The Maximum Likelihood Estimation Method

In fact, even with a toy model like the one described earlier, it is hard to see straightforwardly how Weisberg's account could really apply. This is partly because Weisberg does not give us any specific example about his account's application and partly because its applicability is still in dispute (I argue this in Section 5.5).

In contrast, the toy model can be readily treated by many scientific testing methods, and *maximum likelihood estimation* (MLE) is a very common one[8] I elaborate in this section.[9] In preparation, let me first briefly outline the basics of this method and then use this method to test a more complicated model we introduced in the last chapter, namely, the leaf gas-exchange model. The model and the following description of the MLE method are paraphrased from Shipley (2002, 103–135) with minor modifications.

The essence of the MLE method lies in the comparison between the *predicted* and the *observed* variance and covariance matrices. Specifically, we first specify the causal structure of the model under investigation and then translate it into a statistical model in the form of structural equation(s).[10] After this, we use certain covariance algebras to derive the *predicted* variance and/or covariance[11] between variables in the model. The covariance of variables A and B is expressed as $Cov(A,B)$. Note that the *predicted* variance and/or covariance is derived purely from the structural equations without appealing to empirical data. If we have also collected empirical data about the target system, we can derive the *observed* variance and/or covariance between variables.

The next step is to obtain the MLEs of the free parameter(s)[12] of the model by minimizing the difference between the observed and predicted variances and/or covariances.[13] The typical way to do this is to choose values for parameters in the predicted covariance so as to make it as numerically close as possible to the observed covariance. The process of choosing values for parameters is roughly like searching for the lowest point in a landscape with many hills and valleys. Suppose a lady is blindfolded and her business is to find the lowest point in the landscape without peeking:

> She begins by taking an initial step in a direction based on her best guess. If she sees that she has moved down-slope then she continues in the same direction with a second step in the same direction. If not, she changes direction and tries again. She continues with this process until she finds herself at a position on the landscape in which every possible change in direction results in movement up-slope. She therefore knows that she is in a valley. Unfortunately, if the landscape is very complicated she may have found herself in a small depression rather than on the true valley floor. The only way to find out would be to start over at a different initial position and see whether she again ends up in the same place.
>
> (ibid., 111)

After fixing the values of parameters, we then calculate the number of degrees of freedom in the model, which is given by the function: $v(v + 1)/2 - (p + q)$, where v is the number of variables, "q is the number of free variances of exogenous variables (including the error variables) in the model and p is the number of free path coefficients in the model" (ibid., 115).[14] Finally, based on the values of the parameters and the number of degrees of freedom, we are able to calculate the probability of having observed the measured minimum difference between the predicted and observed covariance[15] and conclude whether our model is good enough in terms of a certain criterion (say, above 0.05 is good enough).[16]

Now, let us consider a more complicated model and its testing procedure using the method sketched earlier. Shipley and Lechowicz (2000) propose a directed path model[17] about leaf gas-exchange based on five variables: specific leaf mass (SLM: leaf dry mass divided by leaf area, g/m^2), leaf organic nitrogen concentration (mmol/m^2), stomatal conductance to water (mmol/m^2 per s), net photosynthetic rate (μmol/m^2 per s) and internal CO_2 concentration (μl/l). (There are reasons why we choose this model but not others. First, it is definitely a mathematical model and falls into the category of minimal modeling according to Weisberg's terminology. Second, it is neither too complicated nor too simple to grasp. Third, and most important, it is a good exemplar in the discourse of causal modeling and related testing methods.) The model is expressed in Figure 5.2.

In Figure 5.2, directed lines represent causal relations, and square boxes represent causal relata. a_x and b_y are free parameters, and ε_n denotes an error variable representing other unmodeled causes of the variable.

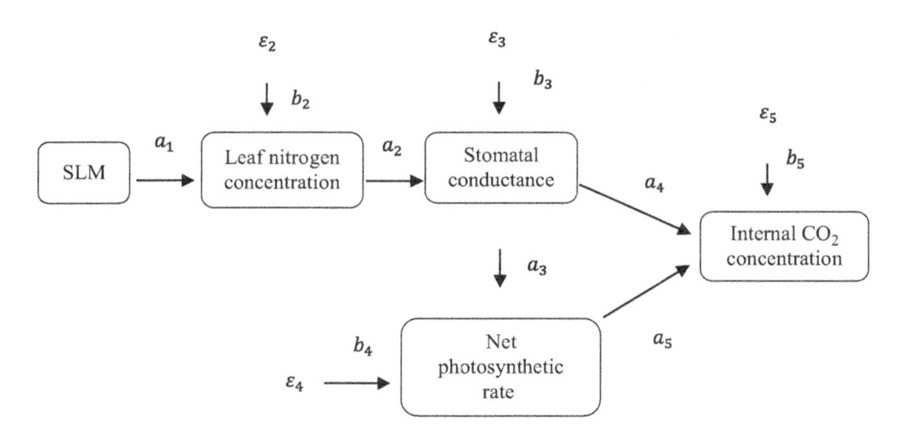

Figure 5.2 The proposed path model relating leaf morphology and leaf gas exchange. The letters with subscripts show the free parameters whose maximum likelihood estimates must be obtained.

Sources: The figure is drawn from Shipley (2002, 131), with only small modifications. Figure used with permission.

Now suppose we intend to evaluate whether this model is scientifically acceptable. Having specified the causal structure, the next step is to translate it into an observational (i.e., statistical) model in the form of a set of structural equations (for technical details, see Appendix 1, Box 1).[18] The third step is to derive the predicted variance and covariance between each pair of variables in the model using covariance algebras. Following the algebras, we finally obtain a table showing the variances and covariances between each pair of variables (see Appendix 1, Box 2).

The fourth step is to estimate the free parameters by minimizing the difference between the observed and predicted variances and covariances. As described earlier, the typical way to do this is roughly like searching for the lowest point in a landscape with many hills and valleys. More specifically,

> [w]e start with an initial guess of the values of the free parameters and calculate the likelihood of the data given the current parameter values. We then see whether we can modify our guess of the values of the free parameters in such a way as to improve the likelihood. We continue with this process until we find values such that any change to them will do worse than the present values.
>
> (Shipley 2002, 110–111)

Fortunately, many computer programs can perform this somewhat time-consuming process,[19] so for simplicity, we here just follow the final outcome of this process. The final maximum likelihood value reached is 4.72, the magnitude of which indicates the difference between the predicted and observed covariance matrices.[20] In general, the larger the maximum likelihood value, the further the predicted covariance matrix deviates from the observed covariance matrix (ibid., 119). This value matters to the following calculation of the probability of having observed the measured minimum difference between the two matrices. That is, to evaluate the model, we first need to make the null hypothesis that the model is correct and then "calculate the probability, based on this null hypothesis, of observing at least as large a difference between the observed and predicted covariance matrices as measured by our statistic" (ibid., 119).

Finally, to determine whether the model fits the empirical data well, we have to calculate the probability of having observed the measured minimum difference, assuming that the observed and predicted covariances are identical except for random sampling variation.[21] To achieve this, we first calculate the degree of freedom in the model. In our model, we have five variables, five error variables, and five free parameters, so the degree of freedom is 5. Given the final maximum likelihood value (i.e., 4.72) and the degrees of freedom (i.e., 5), we arrive at the conclusion that the probability of having observed the measured minimum difference between the predicted and observed covariance matrices is 0.45.[22] According to one commonly used interpretation in biological modeling, if the calculated probability is small enough (e.g., below 0.05), then one can conclude that the model is wrong, whereas if the probability is big enough (e.g., above 0.05), then one can conclude that the model is sufficiently consistent with the data (or not inconsistent enough to reject) and thus is acceptable (Shipley 2002, 115–116).[23] Therefore, since our probability is sufficiently larger than 0.05, we conclude that the leaf gas-exchange model is good enough.

5.5 Can Weisberg's Account Capture How the Gas-Exchange Model Is Assessed?

We mentioned earlier that Weisberg's account attempts not just to generally explain *what* the representational relation between models and the world is but, more important, to also capture similarity judgments made by scientists. Our question hence is to what extent Weisberg's account, intended to capture similarity judgments made by scientists, could *capture* the relevant judgment involved in the leaf gas-exchange modeling discussed earlier.

To answer this question, we first need to identify which aspect of how the leaf gas-exchange model was assessed involves a similarity judgment. Given that the MLE method seems to involve a *comparison* of the predicted with the observed data, let us suppose, for Weisberg's argument, that MLE can be thought of as a sort of similarity judgment. However, we shall see in the following that even if MLE is understood as a sort of similarity judgment, Weisberg's account still fails to capture that judgment. In what follows, I first raise a less important concern and then move on to develop two crucial, interconnected arguments.

5.5.1 Weisberg's Account Is Too Abstract

My first concern is that his account is too abstract to shed light on how the practice of selecting and weighting features (i.e., the "move-back-and-forth practice") can be actually made by scientists.

Needless to say, it is a basic practice in modeling that scientists move back and forth to adjust the features based on their intuitions about the similarity and/or change the model based on similarity values achieved. But the trouble is that this practice is not instantiated in Weisberg's account, for his account only claims that scientists can decide which features should be included in the similarity calculation and how, importantly, they should be weighted, respectively, given certain background theory or theories. He does mention that when the background theory is not rich enough, deciding which features should be weighted more heavily is in part an empirical problem. But the problem remains: How could his account cash out this "move-back-and-forth" issue when the background theory is not rich enough? No clear image emerges. Nevertheless, the empirical issue of selecting and weighting features stands at the core of modeling practice. Hence, as an account of capturing similarity judgments made by scientists, merely mentioning that "this is an empirical problem" falls short of casting light on what is really going on in science.

By contrast, the "move-back-and-forth" practice is perfectly exemplified by MLE. We concluded in the previous section that the probability of having observed the measured minimum difference is 0.45, which is sufficiently larger than the threshold value 0.05. Hence, we claimed that our model is good enough. If, on the other hand, the probability had not been 0.45 but rather 0.045, smaller than the threshold value, we would have concluded instead that our model was not good enough and that there must be features the model missed or misrepresented. If this had been the case, we would have modified the model slightly by adding or subtracting more features to/from the model. We would have practiced this again and again until we reached an acceptable probability value.

In summary, because of its abstractness, Weisberg's account simply fails to shed light on how the process of selecting and weighting features can be actually exercised by scientists. This might not constitute a serious problem because, he might argue, the practice could be integrated into his account someday. I agree, but two serious problems make this integration unpromising.

5.5.2 Weisberg's Account Is Atomistic in Nature

The first serious problem is that his account is *atomistic* in spirit, while MLE is *holistic*.[24] To show this, let us take a closer look at his similarity equation (Equation 5.1). The numerator invites us to weight shared features, and the denominator asks us to weight all features involved (including three feature subsets: shared features, features possessed by the model but not the target and features possessed by the target but not the model). Each feature is weighted independently and only once, with it falling into one of the three feature subsets. The numerator is the weighted sum of shared features, the denominator is the weighted sum of shared and unshared features, and the similarity measure is the ratio of the numerator to the denominator.

But it is evident that this is not what we find in MLE. Consider the leaf gas-exchange model: when testing this model, we are interested in exploring whether the predicted covariance matrix of the model *as a whole* fits the observed covariance matrix using the MLE method, rather than weighting whether each individual feature, or each individual covariance predicted by the model, *matches* each individual feature or covariance of the observed data (remember that in the testing, we choose values for free parameters for the whole predicted covariance matrix and see how the matrix as a whole fits the observed matrix). In other words, the final measure of fit concerns not simply the sum of each feature's measure of weight or the ratio between the sums of weights of the numerator and the denominator. Put slightly differently, the essence of the testing, at least in our model, is not based on weighting each feature independently and then adding them together but on the *holistic* relationship between the predicted data and the observed data *as a whole*.

Three possible responses are on offer. First, Weisberg might launch a rescue by his notion of *fidelity criteria*. In fact, Weisberg does discuss the fit between the output (i.e., predictions) or internal causal structure of the model, on the one hand, and their counterparts in the real-world phenomenon, on the other. For example, he says that dynamical fidelity criteria[25] tell us how close the output of the model must be to the output of the real world phenomenon.... These criteria deal only with the output of the model, that is, its predictions about how a real-world phenomenon will behave (Weisberg 2013, 41). And the representational fidelity criteria "specify how closely the model's internal structure must match the causal structure of the real-world phenomenon to be considered an adequate representation" (ibid., 41). When discussing the representational ideal of MAXOUT,[26] he also says that "this ideal says that the theorist should maximize the precision and accuracy of the model's output" (ibid., 109). These points suggest that sometimes to judge whether a model is adequate, modelers primarily pay attention to the overall fit of outputs or causal structures of the model to the target. That is, they focus on how *similar* the output or causal structure of the model is to their real-world phenomenon. Interestingly, this is just what the MLE method implies, for it concerns the overall fit between the predicted and observed covariance matrices.

But, unfortunately, there seems no way to integrate these ideas into Weisberg's account of similarity between the model and the target. His formula simply does not (and perhaps cannot) incorporate a *holistic* comparison of either outputs or causal structures between the model and the target. The formula essentially depends on assessing the (weighted) contribution of each feature independently. This is largely due to his understanding of the structures of models, a crucial point we shall consider in our last argument.[27]

The second possible response may be that Weisberg's similarity account may be compatible with our holistic diagnostic. Suppose, for example, that the MLE method reveals that the model for leaf gas exchange is wrong. Presumably it is wrong because modelers in constructing it incorrectly identified or weighted certain features of the model. Fixing the model to make it sufficiently similar to its target might require adding or subtracting features *separately*, or weighting them differently. In doing so, goes the argument, an atomistic approach to similarity is employed. But, as we shall see in the last argument, since each individual feature of the model or target cannot be weighted independently without affecting the weighing of other features, adding or subtracting features (or weighting them differently) cannot keep other features (and perhaps relations between them) intact; that is, it must result in the reweighting of many other features and perhaps (if it is a causal model) the reweighting of many relations between these features. Again, this is clearly a holistic issue.

The last possible response might involve the distinction between different representational strategies,[28] or representational ideals in Weisberg's terminology.[29] Weisberg might rightly point out that the holistic character of MLE presupposes the upholding of the representational strategy of *precision* (Weisberg's own terminology is MAXOUT; see Note 26), whereby modelers are only interested in the precision of predicted outcomes, although unrealistic assumptions are always made. This is true in MLE, where scientists make a number of unrealistic assumptions, for instance, that the causal relations involved are additively linear, that all error variables are unit normal variables, and so on. Despite these unrealistic assumptions, the predictions based on MLE can be made very precise.[30]

This is also true in the modeling practice of many other domains, economics, for example. Alan Musgrave (1981) says that [e]conomic theorists often make assumptions which seem to be quite obviously false. For example, an economist may assume that goods are infinitely divisible, that consumers have a perfect knowledge of them, that transport costs are nil, that the government's budget is balanced, even that there is no government (377).

This basic fact aroused the famous *Friedman twist* in economics, where Friedman claimed that "an economic theory should not be criticized for containing 'unreal assumptions': the only legitimate way to criticize an economic theory is to point out that its predictions are at variance with the facts" (ibid., 377).

Thus, the representational strategy of precision is in itself unproblematic. But Weisberg might argue, the worry is that this strategy is just one among many, and focusing solely on it might stretch it too far. Admittedly, our considerations discussed so far might just amount to tilting at windmills, for perhaps his account holds an entirely different representational strategy, *realism*, for example. Indeed, it should be accepted that his account bears a closer relationship to the strategy of realism than to generality and precision (or other strategies), for it turns heavily on the one-by-one manner of feature matching. Note that his account requires that a model be *maximally similar* to its target in important respects. But if Weisberg's objection proceeds in this way, its functioning as a general account of the model–world relationship might be greatly undermined, for it might turn out that it only covers a small portion of modeling practice where the representational strategy of realism is employed, and in particular, it fails to cover MLE in which precision is deployed. This result is not itself too bad since covering a bit is better than covering nothing. But we shall see in the next argument, even the hope of maintaining this small portion is at risk.

In summary, it turns out that the similarity account is *atomistic* in nature while the "similarity judgment" in MLE is made in a *holistic* manner. Further, his considerations about "fidelity criteria" offer no help. Weisberg might rescue his account by claiming either that his account is compatible with MLE, or that his account exemplifies the representational strategy of realism while MLE the strategy of precision. But neither is convincing, for the former is simply untrue and the latter defeats itself. Perhaps it is not an accident that his account shows an atomistic flavor, the reason for which I explore in the next argument.

5.5.3 Weisberg's Account Assumes a Set-Theoretic Approach

The second concern discussed earlier can be traced back to a deeper one, that is, Weisberg's understanding of the structure of models. Weisberg's (2013, 15) basic claim is that models are *interpreted structures*. Corresponding to his three different kinds of models, Weisberg distinguishes three distinct structures: concrete, mathematical, and computational structures. Of direct relevance to the discussion here is the mathematical structure, for the leaf gas-exchange model clearly falls into the category of mathematical models. He states that in many cases of modeling these mathematical structures are *trajectories in state spaces*, but in other cases, other kinds of mathematical structures can also be deployed (2013, 29). It is worth noting here that the trajectory (or state-space) view of structures of models stems from the semantic view of models.[31]

Interestingly, although Weisberg (2013, 137–142) rejects the semantic view of the model–world relation, his similarity account of the *model–world relation* seems to imply a set-theoretic approach that resembles the one found in the semantic view.[32] Given that the adequacy of the semantic view on the nature (i.e., structures) of models has been challenged by many authors (Downes 1992; Frigg 2006; Godfrey-Smith 2006; Odenbaugh 2008), and my argument in this chapter concentrates solely on the model–world relation rather than on the nature of models, in what follows I only consider the point that Weisberg's similarity account on the model-world relation seems to imply an inappropriate set-theoretic approach. Let us demonstrate this.

It can be proved that Weisberg's similarity measure is equivalent to the *Jaccard similarity coefficient* between two sets (see Box 5.1).[33] Box 5.1 should make it intuitively clear that Weisberg's account says just what the Jaccard similarity coefficient says—that the degree of similarity between two sets of things depends on the ratio of elements shared to the total elements of these two sets of things. The full demonstration can be found in Appendix 2.

Box 5.1 The Jaccard Similarity Coefficient

The Jaccard similarity coefficient is a statistical tool that can be used to measure similarity, dissimilarity, or distance between sample sets. For two nonempty finite sets X and Y, the Jaccard coefficient is the ratio of the number of elements in the intersection $X \cap Y$ to the number of elements in the union $X \cup Y$:

$J(X, Y) = |X \cap Y|/|X \cup Y|$ (If X and Y are both empty, we define $J(X, Y) = 1$.)

It measures the probability that an element of at least one of two sets is an element of both and is thus a reasonable measure of similarity or "overlap" between the two (Levandowsky and Winter 1971, 34).

Therefore, it seems that Weisberg's approach to similarity assumes a *set-theoretic* way of describing objects (models and targets) in which features are independent of each other, just as elements of a set are independent of each other.[34] A set-theoretic way of describing objects is one that views both the model and the target as a set of independent elements, the similarity between which consists in the ratio of the number of elements shared to the number of the total elements. However, as my second argument (and MLE practice) shows, modelers, at least in many forms of mathematical modeling, do not engage in set-theoretic descriptions and comparisons of their models with their targets. Rather, they judge their models as wholes. This is because models are *organized structures*,[35] in which features bear strong relationships (in contrast with elements of a set) to one another, rather than *aggregative structures*, in which features are independent. The difference between the set-theoretic and the non-set-theoretic ways of describing objects can be brought out by a simple figure.

Figure 5.3 reveals that if features are interconnected with one another, then it is simply misleading to assess the fit between the model and the target in a set-theoretic way, for this inevitably severs the interconnections among features.[36] This is shown in the next chapter, where we will see that, by taking a closer look at the testing procedure of the leaf gas-exchange model, no one feature in the model can be weighted independently of the other features.

The difference (between the set-theoretic and non-set-theoretic approaches) both echoes and further explains my second concern regarding Weisberg's account, in which I claimed that his account is *atomistic*,

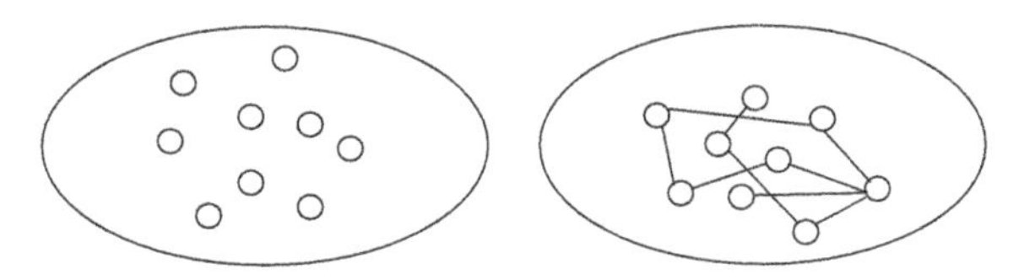

Figure 5.3 Set-theoretic and non-set-theoretic ways of describing objects. The former assumes that features of the object are independent while the latter assumes that they are interconnected.

while MLE is *holistic*. I hope it now becomes clear that this difference has a deep root in distinct understandings of the structure of models. For if we embrace the set-theoretic approach that can be derived from the Jaccard similarity index (as we do in Appendix 2) and thereby assume that features in models are independent of each other, then an atomistic account ensues. Obviously, an atomistic understanding of features precisely corresponds to, and can be explained by, a set-theoretic conception about structures.

This difference also partly explains why Weisberg (2013, 155) rejects the holistic approach to similarity; he states that his "notion of similarity begins from an everyday notion, but rejects the idea that similarity is a strictly holistic relation of resemblance". For if similarity is a holistic relation of resemblance, similarity judgments cannot be simply made in a set-theoretic way. In other words, in order to maintain his set-theoretic account, a holistic approach must be rejected. This does not mean that no similarity account can be made consistent with the holistic approach in the end, only that Weisberg's version of similarity account is as a matter of fact inconsistent with the holistic approach.

So, as I suggested in the last section, the possibility of tailoring Weisberg's account to cover holistic model assessment can be blocked, and even the hope of maintaining the similarity account for the small portion of modeling practice involving the representational strategy of realism is at risk (for its assumption of the set-theoretic approach makes it inconsistent with most—if not all—mathematical modeling).

In sum, given these considerations, I am inclined to conclude that Weisberg's account fails to capture those similarity judgments made by scientists, for these judgments are simply not what his account suggests. Its failure might at first blush be due to its atomistic conception of features, but the underlying bedrock might be due to its set-theoretic conception about the structures of models. Its failure, however, might hint at a new direction in which an alternative approach can be proposed. This is what I describe in the next chapter.

5.6 Conclusion

In this chapter, I have argued that Weisberg's *weighted feature-matching* account of similarity is not as plausible as it seems. To achieve this, I first examined how to use MLE to evaluate the goodness-of-fit of the leaf gas-exchange model. After that, I considered the problem of whether Weisberg's account could accommodate this case study. We saw that it fails as an account aimed at capturing similarity judgments made by scientists. First, his account is simply too abstract. Second and more seriously, it implies an atomistic conception of features while our case treats features holistically. Third and most important, the atomistic conception of features can be traced back to a set-theoretic view of the structures of models, where elements of a set are independent of each other.

Notes

1 Wendy Parker expresses a similar interpretation of Weisberg's account, claiming that "[u]ltimately, however, Weisberg's account seems best characterized as an account of what underwrites scientists' *judgments* of the *extent* to which models and targets are similar" (2015, 271; author's emphasis).
2 Weisberg's account can be more strongly interpreted as aiming to "capture *how* similarity judgments are made by scientists", for his similarity equation and its associated interpretation (as I outline later) includes the way in which features are *selected* and *weighted* differently by scientists. This clearly involves the "how" problem: How do scientists select and weight these features so as to make appropriate similarity judgments? But to be charitable, in this chapter, I simply interpret his account as attempting to "capture similarity judgments made by scientists". The latter interpretation is weaker than the former because it merely interprets Weisberg's account as *describing* similarity judgments rather than the process by which they are made.
3 Weisberg also describes three other kinds of modeling practice that require different weighting parameters: hyperaccurate, how-possibly and mechanistic modeling. For details, see Weisberg (2013, 150–152).
4 Let us assume that there is such a data set regarding the variables in the model. Note that this extremely simple case should be regarded as a textbook example used merely for introductory purposes.
5 This model is borrowed from Shipley (2002, 47–48), with minor modifications.
6 I argue extensively in Section 5.5 that the applicability of Weisberg's account is, in fact, in dispute.
7 Because we have already assumed that there is a causal mechanism between the two variables and that is just what the modelers aim to capture.
8 There are two common ways of testing causal models, of which the MLE method is the most common one; another way is called *d-separation tests*. For details of the second way, see Shipley (2002, 71–99).
9 Although I do not demonstrate the claim that the toy model can be readily treated by many scientific testing methods, the following elaboration of the MLE method regarding a more complicated model should lend support to this claim.
10 Variables on the right side of the equation are causes and variables on the left side are effects. ε is an error variable representing other unmodeled cause of the variable E, or pure randomness.
11 The variance is the expected square of the deviation around a variable's expected value, defined as $E = [(X_i - \mu_X)]^2$, where μ_X is X_i's expected value. The covariance is simply a generalization of the variance. If we have two different random variables (X, Y) measured on the same observational units,

then the covariance between them is defined as $E[(X_i - \mu_X)(Y_i - \mu_Y)]$, in which $(X_i - \mu_X)$ and $(Y_i - \mu_Y)$ represent deviations of X and Y from its mean μ_X and μ_Y, respectively (Shipley 2002, 74).

12 A free parameter is one that can be adjusted to make the model fit the data.

13 This is typically done by using the maximum likelihood chi-square statistic. For details, see Shipley (2002, 110–114).

14 An exogenous variable in a model is one that does not have causal parent(s) in the model; for example, SLM and ε_i are exogenous variables in the leaf gas-exchange model; in contrast, an endogenous variable in a model is one that does have causal parent(s) in the model. Variable X is a causal parent of variable Y in a model if X is a direct cause of Y in the model, and *direct cause* means there are no intermediate causes between X and Y in the model.

15 The MLEs (typically using the maximum likelihood chi-square statistic) guarantee that, by iteratively choosing values for free parameters, the numerical values of the predicted covariance matrix become as close as possible to the actual covariances measured in the data. A caveat: the iterative procedure could sometimes get stuck by local maxima without finding the global maximum (I thank Arnaud Pocheville for letting me notice this problem). The only way to avoid this problem is to try different starting points and see if they get to the same values (Shipley 2002, 113). In essence, to increase the fit as close as possible is also to decrease the difference as much as possible. Hence, at a certain point, we obtain a minimum difference. The next step is to calculate the probability of having observed such a minimum difference. The value of probability can be calculated by some commercial computer programs, given certain MLEs and degrees of freedom.

16 In scientific testing, the acceptability criterion of 0.05 (or some other values such as 0.01 or 0.1, depending on different fields of study or different questions under consideration) is also called the *significance level* or P *value*, used by scientists to determine whether the null hypothesis should be rejected or not. In short, the significance level indicates whether there is a relationship between various variables represented in the model under testing or whether the result can be explained by the null hypothesis.

17 "A *directed path* between two vertices in a causal graph exists if it is possible to trace an ordered sequence of vertices that must be traversed, when following the direction of the edges (head to tail), in order to travel from the first to the second" (Shipley 2002, 27; author's emphasis). A vertex in a model represents a variable in that model.

18 In most structural equations, the causal relations are assumed to be additively linear (Shipley 2002, 105), so we here follow this assumption. It should also be noted that since causal relations are asymmetric while these equations are symmetric, this translation is only a partial translation.

19 See Shipley (2002, 110–114) for more details about these techniques.

20 This difference can be calculated using some formulae. For more details, see Shipley (2002, 113–114).

21 See Note 15.

22 See Shipley (2002) for more details about the calculations of this value.

23 There might be other interpretations that set different threshold values for the acceptable probability, but our case discussed in this chapter is neutral with exactly which value is actually chosen.

24 Wendy Parker recently made a similar point, asking, "[C]an feature weights really be assigned independently?" She says that "[i]n his account of model-world similarity, Weisberg deviates from Tversky in restricting the weighting function such that it assigns weights to features independently of which

other features are shared" and that "Weisberg's restriction seems to go too far, however. For surely the perceived significance of a feature 'shared' by a model and a target sometimes does depend on which other features are 'shared'" (Parker 2015, 273–274). In response to Parker's criticism, Weisberg (2015) admits that in some cases, "we should see the independence assumption as relaxed, or, perhaps, we could construct compound features of the tightly coupled ones, and weigh them accordingly" (304). However, Weisberg does not state clearly how the independence assumption can be relaxed in these cases. Moreover, the amendment that "constructing compound features of the tightly coupled ones and weighing them accordingly" would move his view in the direction of the holistic view I develop in the following chapters.

25 For Weisberg (2013), fidelity criteria "describe how similar the model must be to the world in order to be considered an adequate representation. There are two types of fidelity criteria: *dynamical fidelity criteria* and *representational fidelity criteria*" (41).

26 MAXOUT is a representational ideal that guarantees that models are useful for predicting but gives no guarantee that the models are useful for explaining (Weisberg 2013, 109).

27 One may argue that there does exist a way to integrate these ideas into Weisberg's account, for example, Weisberg's features may also apply to system-level features, for example, an overall pattern of a system rather than any specific variable in that system. However, although this move sounds sensible, it collapses Weisberg's account into a holistic account I describe in the following chapters. I thank Patrick McGivern for letting me notice this potential response.

28 The discussion surrounding representational strategies started with Richard Levins (1966), who claimed that there are three types of representational strategies (i.e., realism, precision, generality) among which trade-offs exist, and continued well into today with Orzack and Sober (1993), Taylor (2000), Odenbaugh (2003, 2006), Orzack (2005), Weisberg (2006), and Matthewson and Weisberg (2009), among others.

29 Weisberg himself also distinguishes various different representational ideals: completeness, simplicity, 1-causal, maxout and P-general. For details, see his book (Weisberg 2013, Chapter 6, Section 6.2).

30 The next chapter will show that, insofar as MLE practice is not merely associated with predictive models, saying that MLE is only concerned with the representational ideal of prediction accuracy is untrue.

31 The semantic view of models (or more generally about theories) has two versions: one is the set-theoretic predicate approach developed by Suppes (1957, 1960, 1962, 1967), Sneed (1971), and Stegmüller (1976), and the other is the state space approach developed by van Fraassen (1970, 1972, 1974) and Suppe (1974, 1977). The next chapter says more about the semantic view.

32 The semantic view of models, as Weisberg (2013, 137–138) himself notes critically, takes "mathematical models ... to be more or less equivalent to logicians' models" and thus accounts for "the model-world relation using tools which are appropriate for logicians' models". In particular, it takes the model–world relation to be an isomorphic relationship (or some weakened version), that is, a mapping between two *sets* that preserves structure and relations (ibid., 137–138).

33 I thank Arnaud Pocheville for bringing this to my attention.

34 There are two senses of independence to be distinguished in the next chapter. First, in contrast to causal interaction, there exists a kind of causal independence among variables: there are no causal interactions among these

variables. Second, in contrast to similarity interaction, there exists similarity independence among variables, meaning that one variable's contribution to the similarity measure of the model with respect to its target system does not depend on the other variables' contribution to that measure. In this chapter, when assessing Weisberg's account, I mainly refer to the second sense of independence. I thank Patrick McGivern for alerting me to this point.

35 I say more about my conception of model structure in the next chapter.

36 A proponent of similarity view can reply that the whole structure composed of various interconnected elements can be thought of as a single feature, and the similarity judgment is just made about the similarity of this single feature between the model and the target. But this reply makes Weisberg's view descend into a version of a holistic approach that he rejects.

6 A Holistic View of the Model–World Relationship

6.1 Introduction

The last chapter demonstrated that Weisberg's similarity account of the model–world relationship cannot shed light on the leaf gas-exchange modeling. This is due to its atomistic conception of features and its assumption of the set-theoretic approach to model structures. Against such a backdrop, this chapter aims at developing a holistic alternative to Weisberg's view by means of looking closely at scientific practice.

The last chapter gave us a preliminary understanding of the view that the model–world relationship is a holistic relationship. To cash out that view, this chapter suggests that, for a good model,[1] the model–world relationship is a holistic fit, wherein where *holistic fit* means the structure, output, or pattern[2] of the model fits the structure, output, or pattern of the target *as a whole* (for simplicity I only use the term *structure* in what follows, except for places where distinctions among the three are needed). Put differently, a holistic fit refers to the degree to which one structure resembles another structure or, to say it in another way, refers to the *distance* between two structures.[3] More precisely,

Definition: the model–world relationship constitutes a holistic fit when it satisfies two conditions:

a. the condition of causal interaction: the modelers take the model (and its target system) as an interconnected whole, wherein components of the whole interact with one another in producing certain outputs (or patterns, phenomena, etc.), and
b. the condition of similarity interaction: the similarity measure of the model with respect to its target system is achieved via comparing outputs (or patterns, phenomena, etc.) produced by these interacting components. In other words, one feature's (a variable, a relation between variables, etc.) contribution to the similarity measure depends on other features' contributions to that measure (Section 4.1 says more about similarity interaction).

DOI: 10.4324/9781003148029-6

Before proceeding, it might be helpful to clarify one thing concerning model adequacy. This chapter takes it as a background assumption that a modeler's modeling goal (or representational ideal) would shape their modeling practice and that a good model is usually one that is judged against a certain modeling purpose, for example, explanation (or realism), prediction (or accuracy), extrapolation (or generality), and so on (Levins 1966; Orzack 2005; Odenbaugh 2006; Weisberg 2006; Matthewson and Weisberg 2009; Parker 2009, 2010). In other words, there is no good model *simpliciter*, only a good model with respect to its modeling goal. Although this chapter centers on how scientists evaluate the goodness-of-fit of their models and does not address explicitly how the evaluation practice matters to the problem of model adequacy, it goes without saying that the former is closely related to the latter. In particular, an evaluation is usually an evaluation of a predefined model which has a clear modeling goal (and perhaps a clear criterion for assessing how the model meets the specified goal, i.e., assessing the adequacy of the model) when building it. Therefore, it is more precise to say that we are always evaluating a model with particular goals rather than evaluating a model *simpliciter*. This means that, to use Wendy Parker's (2009, 233) terminology, what modelers evaluate is not the model itself but rather the hypotheses about the adequacy of the model for particular purposes. Hence, we can see the adequacy consideration as an embedded component of the evaluation practice—the adequacy consideration is already there when evaluating a predefined model. For this reason and because of space limitations, this chapter does not discuss how a modeling goal would shape one's modeling practice, nor does it discuss how the adequacy consideration could be embedded in one's modeling practice.[4]

The strategy of this chapter is to reexamine maximum likelihood estimation (MLE) practice at length for the purpose of deriving philosophical implications pertaining to fleshing out the holistic view. The layout is as follows: to pave the way for reexamining the somewhat sophisticated MLE practice in Section 6.3, Section 6.2 first considers a very simple estimation method commonly used in practice, namely, the least squares estimation (LSE) method. This is done by considering the curve-fitting problem using a simple biological example. It turns out that scrutinizing LSE leads to the same philosophical implications as those we find in MLE in Section 6.3. Section 6.4 tentatively suggests that, although one of the goals of this book is to develop a holistic view about biological models, the philosophical implications derived from discussing biological models can be further generalized to nonbiological models. For this purpose, a concrete model, that is, the San Francisco Bay model, is discussed. The general conclusion is that, for a good biological model, the model–world relationship can be best viewed as a holistic fit. Finally, given that the notion of structure is also invoked in the semantic view of models, Section 6.5 will try to distance my view from the semantic view by developing a deflationary account of model structures.

6.2 Least Squares Estimation

For various reasons, MLE is a very common estimation method used in scientific practice (Shipley 2002, 71; also see Myung 2003; Kleinbaum and Klein 2010).[5] Thus, insights into models and modeling can be gleaned by taking a closer look at MLE practice. Yet, to fully understand MLE practice, it is better to first consider a simpler estimation method, that is, the LSE method, which is also widely used in scientific practice (e.g., in biology, see Allman and Rhodes 2004; Symonds and Blomberg 2014; in psychology, see Usher and McClelland 2001; in astronomy, see Nievergelt 2000).[6]

For simplicity, I only consider linear LSE in what follows.[7] The following description of the LSE method largely follows Rice (2006), although with many modifications. To a first approximation, LSE is mainly used in curve-fitting problems to estimate parameters by minimizing the sum of squared deviations of the predicted (or fitted) values (given by the curve) from the observed values (Rice 2006, 542). To see how the estimation works, consider a simple example: a biologist is interested in whether the plant's height (y) is related with the amount of fertilizer (x) the plant receives. Suppose she also believes that, if y is related with x, then the relation must be linear. In this case, x is called the predictor variable and y the response variable. The biologist then proposes a model for the two variables:

$$y = \beta_0 + \beta_1 x, \tag{6.1}$$

where β_0 is the intercept and β_1 the slope of the proposed line. Remember that the biologist's purpose is to choose parameters to minimize the sum of squared deviations of the predicted from the observed values, and β_0 and β_1 are the parameters chosen. The following function is the sum of squared deviations:

$$S(\beta_0, \beta_1) = \sum_{i=1}^{n} (y_i - \beta_0 - \beta_1 x_i)^2, \tag{6.2}$$

where n denotes the number of observations and i refers to the ith observation. Now suppose she takes a sample of n plants in the species, observing values y of the response variable and x of the predictor variable. For the moment, the biologist chooses values b_0 and b_1 as estimates of β_0 and β_1, respectively, which are supposed to best fit the sample data. Based on the data collected, the biologist plots a figure for the observed values of x and y:

Note that although there are some variations in Figure 6.1 (e.g., when $x = 5$), a clear trend still can be detected. Next, the biologist defines a fitted function:

$$\hat{y} = b_0 + b_1 x, \quad (6.3)$$

where $\hat{y}$ is the fitted value (in the fitted line, namely, in the straight line in Figure 6.1). Now, for each value of the observed response variable y_i and for each corresponding value of the predictor variable x_i, she has a fitted value $\hat{y}_i = b_0 + b_1 x_i$. As such, it is time to minimize the sum of squared deviations of each observed response from its fitted values. She expands Equation 6.2 to get b_0 and b_1:

$$S(b_0, b_1) = \sum_{i=1}^{n} \left(y_i - \hat{y}_i \right)^2 = \sum_{i=1}^{n} \left(y_i - (b_0 + b_1 x_i) \right)^2. \quad (6.4)$$

It can be shown that b_0 and b_1 satisfy the following formulae (Rice 2006, 544):

$$b_1 = \frac{\sum_{i=1}^{n} (x_i - \bar{x})(y_i - \bar{y})}{\sum_{i=1}^{n} (x_i - \bar{x})^2} \quad (6.5)$$

and

$$b_0 = \bar{y} - \hat{\beta}_1 \bar{x} = \frac{\sum_{i=1}^{n} y_i}{n} - b_1 \frac{\sum_{i=1}^{n} x_i}{n}, \quad (6.6)$$

where $\bar{x}$ refers to the mean of x, $\bar{y}$ to the mean of y and $\hat{\beta}_1$ to mean the estimate of β_1, namely, b_1 (for a discussion of these calculations, see Rice

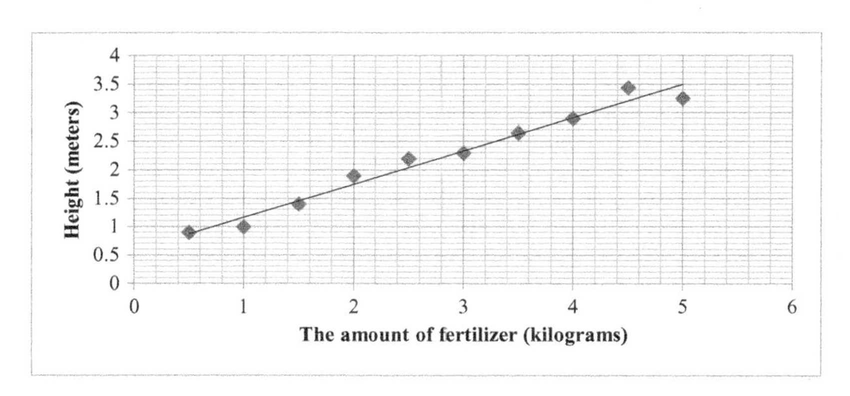

Figure 6.1 The observed data for the height of plants and the amount of fertilizer these plants receive. Note that the data in this example are only made for the purpose of illustration.[8]

2006, 542–547). Finally, after these calculations, the following results are obtained: $b_0 = 0.59$, $b_1 = 0.58$, and therefore, the function for the model is

$$y = 0.59 + 0.58x. \tag{6.7}$$

Having described the estimation method, now let us consider what philosophical implications can be drawn. The first point to note is that we compared two structures by comparing two curves. That is, we compared the fitted curve with the observed curve. This is typically done by minimizing the sum of squared deviations of the fitted values from the observed values. This process is holistic in essence because, when minimizing the sum of squared deviations, we did not care about the behavior of any particular point (e.g., (x_i, y_i)) but did care about the overall effect that is produced interactively by those points' behaviors. More specifically, looking closely at Equation 6.5, we can see that the value of b_1 was achieved by centering the variables x_i and y_i about their means. That is, we first subtracted the values of each variable from their means, and the resultant values were the mean-centered values of the variables. The remaining calculation was accomplished in terms of the mean-centered variables, not in terms of the variables themselves. The mean of a variable describes the central tendency of that variable, namely, the overall behavior of that variable. Besides, we did not consider each variable separately but considered pairs of variables (i.e., (x_i, y_i)) collectively, for example, the numerator in Equation 6.5 was obtained by summing up the product of each pair of the mean-centered values of the two variables (i.e., $\sum_{i=1}^{n}(x_i - \bar{x})(y_i - \bar{y})$. Therefore, we may say that the value of b_1 was achieved by considering the overall behavior of the variables x and y. By the same token, when looking closely at Equation 6.6, it can be seen that the value of b_0 was achieved by considering the means of the variables (i.e., $b_0 = \bar{y} - \hat{\beta}_1 \bar{x}$), not by considering the values of the variables themselves. Arnaud Pocheville (personal communication) has summed up the situation as follows: "Equations 5 and 6 nonlinearly rely on the means of the variables to which all values make a contribution at the same time, hence any independent contribution to the mean is later involved in a network of nonlinearities". I understand this to mean that, when comparing the model with its target, we did so in an essentially holistic way.

One may point out that if there were an abnormal observation point (x_s, y_s) that resides far away from the remaining points,[9] then the resultant fitted curve would be affected by this point. This sometimes does occur, but rather than undermining my stance, it underwrites my viewpoint: since what is under concern is the overall behavior of all observation points, if there exists one (or a few) point that affects the overall behavior of all points, then this point must be taken seriously. For example, we may need to look carefully into the unmodeled cause of the occurrence of this point (e.g., measurement errors or latent variables), and a systematic investigation of the cause may deepen a modeler's understanding of the natural phenomenon in question. As such, we are concerned with the effect of this abnormal point on the overall system's behavior (i.e., the set of points), rather than with the point itself.

The second detail to note is the way we choose the values for the parameters in LSE. As shown above, in LSE we tried to choose the best combination of values (i.e., b_0 and b_1) for the parameters that could minimize the discrepancy between the fitted and observed curve. In other words, we did not really care about the particular value of each parameter; instead, we were caring about the best combination of values for the two parameters that, taken together, could minimize the distance between the fitted and observed curve. Therefore, the act of choosing one value for one parameter would affect the act of choosing values for another parameter. This has been shown by the dependence relationship between calculating values for b_0 and b_1 in Equations 6.5 and 6.6: since $b_0 = \bar{y} - \beta_1 \bar{x} = \bar{y} - b_1\bar{x}$, we may say that the act of choosing values for b_1 would affect the act of choosing values for b_0. In other words, the values for the parameters interact with one another in producing the best combination of values.

In sum, the preceding discussion demonstrated that in scientific practice, we routinely compare the model with its target system holistically. More specifically, we have seen that the LSE practice satisfies the two conditions for a holistic fit (see Section 6.1). Perhaps this conclusion goes too far, because it is based on a single type of modeling practice, that is, LSE. Yet, the following section involving another modeling method, that is, MLE, shows that the conclusion is in fact very general.

6.3 Maximum Likelihood Estimation

The last chapter described the basic steps of employing MLE in testing biological models. There, an intuitive understanding of MLE was sufficient for the purpose of uncovering the shortcomings of Weisberg's similarity view. However, to develop general philosophical implications, we need to look into the practice more closely. This section is devoted to reexamining MLE in depth, with the focus on the fourth step where we estimate the parameters of the model. The following testing procedure follows Shipley (2002) and the relevant mathematical demonstration is also due to him. I try my best to keep to the minimal technical details relevant to developing the holistic view and leave the readers to Shipley's original proof in his book. Finally, since we are interested in what philosophical implications can be derived rather than in the model itself, a simplified leaf gas-exchange model involving only two random and one error variables is introduced in this section.

6.3.1 Presentation of the Method

The following description of the MLE method is paraphrased from Shipley (2002, 103–135) with minor modifications. Recall that by using MLE to

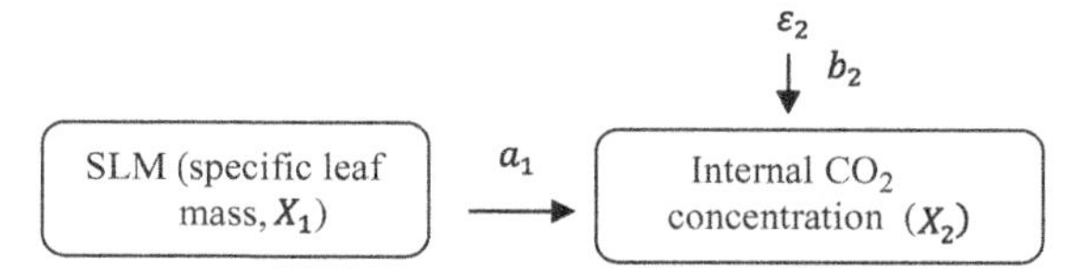

Figure 6.2 A simplified path model relating two random variables.

test the goodness-of-fit of a biological model the first step is to hypothesize the causal structure of the relationships between various variables under consideration. A simplified leaf gas-exchange model is hypothesized as shown in Figure 6.2.

The second step is to translate the causal model into an observational model, and the observational model is usually expressed as a set of structural equations (Shipley 2002, 104–107):

$X_1 = N(0,\Sigma_1)$[10] $\qquad \varepsilon_2 = N(0,\Sigma_2)$

$$X_2 = a_1X_1 + b_2\varepsilon_2$$

$$Cov(X_1,\varepsilon_2) = 0$$

The next step is to obtain the predicted variances and covariances between each pair of variables in the model, and this can be done by using covariance algebra (ibid., 107–110). The result is shown in Table 6.1.

The fourth step is to estimate the free parameters (i.e., a_i, b_j and ε_k) in the model using maximum likelihood estimation. As discussed in Chapter 4, this is done by minimizing the difference between the observed covariances of the variables in the data and the covariances of the variables predicted by the model. The typical way to do this is to choose values for the parameters in the predicted covariance matrix so as to make it as numerically close as possible to the observed covariance matrix. This process is roughly like searching for the lowest point in a landscape with many hills and valleys (as described in Chapter 5). The previous chapter only gave a rough idea of the choosing process; now, let us delve into it.

Let me start with a couple of technical terms. For a random variable, X, if it is continuous, then we need the probability density function. The probability density function of a univariate normal random variable is

Table 6.1 Predicted population variances and covariances for variables

	X_1	X_2
X_1	$Var(X_1)$	$a_1 Var(X_1)$
X_2		$Var(X_2)$

denoted $f(X|\mu,\sigma)$, where μ refers to the mean of the observations of X, σ refers to the standard deviation of X, and μ and σ are population parameters that are fixed.[11] This probability density function says that, given the values of the parameters μ and σ, the probability of X taking on a specific value is $f(X|\mu,\sigma)$. However, more often than not, we do not know the values of the parameters in question (e.g., μ and σ), and it is relatively easier to obtain the values for X by observation (or experiment). Because of this, our goal now is shifted to estimate the values for the parameters given in the observation. In doing so, we obtain a likelihood function $L(\mu,\sigma|X)$, meaning that, given the observation X, the specific values of μ, and σ would maximize the likelihood that we have observed X.

Now suppose we use the likelihood function $L(\mu,\Sigma|X)$ to estimate the values for our parameters under consideration, where Σ refers to the *population covariance matrix*, of which I will say more in what follows.[12] Since what we are interested in are not the exact values of the variables but their relationships, we often center the variables about their means. That is, we subtract the values of a variable from its mean, and the resultant mean-centered variable has a mean of zero. Since the data (i.e., X) and the mean-centered variables are fixed (i.e., their means $\mu = 0$), the only parameters whose likelihood estimates we have to estimate are those in the population covariance matrix Σ. In the simplified leaf gas-exchange model, the matrix Σ is the one that has been described in Table 6.1. The logic of estimating Σ is, as said above, like searching for the lowest point in a landscape with many hills and valleys. More precisely, we first

> group all these free parameters together in a vector called θ. Now, if we take a first guess at the values of these free parameters then we can calculate the predicted covariance matrix based on these initial values; let's call the predicted covariance matrix that results from this guess $\Sigma_{(1)}(\theta)$ to emphasise that this matrix will change if we change our values in θ.
>
> (Shipley 2002, 113)

Given this first matrix $\Sigma_{(1)}(\theta)$, we then calculate the value of the likelihood function $L(\mu,\Sigma_{(1)}(\theta)|X)$. After this, "we change our initial estimates of the free parameters and recalculate the predicted covariance matrix, $\Sigma_{(2)}(\theta)$" (ibid., 113). Based on this new matrix, we calculate the new value of the likelihood function $L(\mu,\Sigma_{(2)}(\theta)|X)$. We repeat this process again and again until we cannot increase the value of the likelihood function. At the point where we cannot increase the value of the likelihood function anymore, we obtain the most likely values for those free parameters in the model.

On the other hand, we can calculate the observed covariance matrix S purely based on the data. With the predicted covariance matrix $\Sigma(\theta)$ and the observed covariance matrix S at hand, the difference between the two matrices can be calculated using the maximum likelihood fitting function F_{ML}.[13] Let us focus on the basic ideas of this function. Recall that our goal is to choose values for the free parameters in the predicted covariance matrix so as to make it as numerically close as possible to the observed covariance matrix. So far, we have chosen the most likely values for the free parameters in the predicted covariance matrix $\Sigma(\theta)$ by the previous step, and the values of the free parameters that maximize the likelihood function $L(\mu,\Sigma|X)$ are also the values that minimize the maximum likelihood fitting function F_{ML}. Namely, these are the values that minimize the difference between the predicted covariance matrix and the observed covariance matrix.

The foregoing process is the fourth step, namely, estimating the values of the free parameters in the model using the maximum likelihood estimation method. The fifth step is to calculate the probability of having observed the measured minimum difference (assuming that the predicted and observed covariances are identical except for random sampling variation). This step involves calculating the degrees of freedom,[14] and testing the null hypothesis that there is no difference between the predicted and observed covariance matrices except for random sampling variation of N independent observations.[15] Suppose that the probability of having observed the measured minimum difference we finally obtain is 0.36, which is higher than the significance level 0.05;[16] that is, our null hypothesis is not rejected by the evidence (i.e., data). In other words, since we do not have sufficient reason to reject the null hypothesis, the null hypothesis is tentatively accepted. If this is true, then our hypothesized causal model, as shown in Figure 6.2, is therefore tentatively accepted.

6.3.2 Discussion: Philosophical Implications

By carefully tracking the estimation procedure, the central implication we get from the foregoing discussion is that, as we have gleaned from LSE practice, the model (the causal graph) fits its target system holistically. Since we cannot directly compare the causal relationships—as represented in the graph—with their counterparts in the real plant system, we instead compared two covariance matrices, one derived from the hypothesized causal graph and one from the real system, respectively. A covariance matrix, to put it metaphorically, is a *shadow* of the model or the target system, which portrays correlations between variables.[17]

Because we started with the hypothesis that it is a causal model, and instead of creating a new model based on data we tested a model against data, the covariances in the matrix (Table 6.1) were hypothesized to measure not only the (statistical) correlations between variables but also the degree to which these variables are causally linked. Grouping these covariances together results in a covariance matrix, which hints at a causal structure within which elements are causally linked to one another directly or indirectly (and, of course, there are always elements in the structure that are not causally linked, e.g., variables X_1 and ε_2 in Figure 6.2). As a consequence, when comparing two covariance matrices, we are, in fact, comparing two (causal) structures, although we do this in an indirect way (as I said, we cannot compare the structure of the model with the structure of the real plant system directly).

Another way to see that the model–world relationship is a holistic relationship is to notice the focus on the relationships between variables rather than on the variables themselves. As in the case of LSE, where we did not care about any particular point's behavior but about the overall effect that is produced interactively by all those points' behaviors, in MLE, we also did not care about the particular value (or behavior) of each variable. Remember that in MLE, we centered each variable around its mean. What we really cared about were the degrees to which one variable would affect another variable, and these influences were expressed by the free parameters (recall the free parameters a_i and b_j in our hypothesized model). By estimating the values of the free parameters between variables, we are measuring the degree to which one variable's behaviors might be affecting another variable's behaviors (i.e., we are therefore measuring the covariances between variables). By estimating the values of all the parameters *at the same time* given the data—recall the process of choosing values for the parameters described earlier—we are, in fact, measuring the whole causal structure; that is, we are measuring the degree to which our model fits the data.

The last but not the least matter to take notice of when considering the model–world relationship to be a holistic relationship is to take a closer look at the estimation of the free parameters. As we have seen in LSE practice, where we tried to choose the best combination of values for the parameters that could minimize the discrepancy between the fitted and observed curves, in MLE practice, the same story happened. Recall the process of choosing values for the free parameters using the maximum likelihood function $L(\mu, \Sigma | X)$. Our goal was to find a set of values for our free parameters that could maximize the maximum likelihood function. The set of values that could maximize the maximum likelihood function were grouped into a vector called θ. Notice that we did not really care about any particular value in the vector; instead, we were caring about the best combination of values for the vector that, taken together, could maximize the maximum likelihood function. Therefore, the act of choosing one value for one free parameter would affect the act of choosing values for other free parameters in the vector. In other words, the free parameters in the vector interact with one another in producing the best combination that could maximize the maximum likelihood function. To give you a vivid image of how the free parameters in the vector might interact with one another in producing the best combination, and, more generally, how the process of choosing values for parameters is holistic in nature, please refer to Appendix 3, where the computer program, Tetrad, is used to evaluate a scientific model based on a hypothetical data set.

Before concluding this subsection, it is worth mentioning one more thing: holistic fit admits both thresholds and degrees. For example, with respect to the same target system, we may build two (or even more) equivalent models that both holistically fit their target, although the degree to which they holistically fit their target can be different.[18] On the other hand, some models may be rejected because of their poor holistic fit towards their target systems. Fixing the degree (i.e., the threshold) to which a model sufficiently holistically fits its target system is a practical problem, which can vary from case to case. However, we should not fuss too much over fixing the exact threshold for a holistic fit, since we can simply follow the logic of testing a model (e.g., using the MLE method), concluding that the model does not sufficiently holistically fit its target if the testing result shows the model is rejected or that the model does sufficiently holistically fit its target if the testing shows the model is not rejected. For example, given the common practice wherein scientists set the significance level for rejecting a null hypothesis at the value 0.05, we may say that ensuring the minimum level for holding an acceptable degree of holistic fit should follow the same logic.

In sum, so far, we have seen from several angles that a model bears a holistic relationship to its target system. First, when comparing a model with its target we are comparing two structures through comparing their covariance matrices (since we often cannot directly compare two structures). Second, the way we compare two structures is via focusing on relationships among variables rather than on variables themselves. Third, the relationships among variables, expressed as free parameters, interact with one another in producing the best fit of the model to its target system (i.e., they interact with one another in maximizing the likelihood function $L(\mu,\Sigma|X)$ and minimizing the maximum likelihood fitting function F_{ML}). Therefore, it satisfies the criterion for a holistic fit (see Section 6.1).

Given the same philosophical implications obtained in LSE and MLE practice, the more general lesson is that in the biological modeling practice, insofar as the model–world relationship is concerned, we usually consider two structures holistically.

6.4 Generalizing the View: The San Francisco Bay Model

The foregoing discussion has shown that biological models bear a holistic fit to their target systems and that this conclusion is insensitive to which particular estimation method is employed. Yet, one might be interested in exploring whether the philosophical implications obtained by scrutinizing biological models are idiosyncratic to biological models or can be further generalized to other kinds of models, nonbiological models for example. If it turns out that the implications hold in nonbiological models, then the holistic view gets extra support with respect to its viability as well as its scope. To this end, this section examines a concrete model, that is, the San Francisco Bay model introduced in Chapter 4.

The first subsection shows that the features in the Bay model interact with one another in producing outputs, and on the basis of this, the second subsection suggests that the Bay model also bears a holistic relationship to its target system.

6.4.1 Features Interact With One Another

As a first step, consider the idea that the Bay model, as a concrete object, is a structure in which features[19] interact with one another in producing phenomena (or outputs). In fact, this is a plain fact for many modelers, for instance, Huggins and Schultz (1973) put it explicitly when testing the bay model:

> Among the problems to be considered were the conservation of water and the preservation of its quality; … the intrusion of salinity into the Sacramento-San Joaquin Delta; the tides, currents and salinity of the Bay as they affect other problems.… *None of these problems can be studied separately, for each affects the others.*
>
> (12; my emphasis)

Consider, for instance, the relationship between two key features in the model: tide and salinity. Salinity changes along an estuary due to the influence of the combination of freshwater and saltwater. An estuary

> is the transition between a river and a sea. There are two main drivers: the river that discharges fresh water into the estuary and the sea that fills the estuary with salty water, on the rhythm of the tide. The salinity of the estuary water is the result of the balance between two opposing fluxes: a tide-driven saltwater flux that penetrates the estuary through mixing, and a freshwater flux that flushes the saltwater back.
>
> (Savenije 2005, Preface ix)

To illustrate this *rhythm of the tide*, consider the effect of the spring-neap tidal cycle on the vertical salinity structure of the James, York, and Rappahannock Rivers, Virginia, US:

> Analysis of salinity data from the lower York and Rappahannock Rivers (Virginia, U.S.A.) for 1974 revealed that both of these estuaries oscillated between conditions of considerable vertical salinity stratification and homogeneity on a cycle that was closely correlated with the spring-neap tidal cycle, i.e. homogeneity was most highly

> developed about 4 days after sufficiently high spring tides while stratification was most highly developed during the intervening period.
> (Haas 1977, 485)

This short report shows not only that characteristics of salinity (such as stratification and homogeneity) are influenced by characteristics of the tide but also that there is a phase connection (or synchronization) between tidal cycle and salinity oscillations. The former is a causal relationship while the latter is a temporal relationship (or a statistical correlation). The phase connection among features was also emphasized by the Army Corps (1963) when verifying the Bay model, saying,

> These gages were installed in the prototype and placed in operation several months in advance of the date selected to collect the primary tidal current and salinity data required for model verification, since *it was essential to obtain all data simultaneously for a given tide over at least one complete tidal cycle of 24.8 hours.*
> (50; my emphasis)

Moreover, the same story goes for tide and tidal currents (for details, see Army Corps 1963, 20).

In short, the features in the model bear not only causal relationships but also temporal relationships. This implies that, when verifying the model, the features of the model causally interact with each other in producing certain outputs (e.g., predictions, effects, phenomena, etc.) rather than individually or separately producing outputs. So although the outputs of the key features in the Bay model can be identified and measured separately, they are not produced separately.

Causal interaction among the features leads us to notice a second kind of interaction, that is, a *similarity interaction*,[20] wherein different features interact with one another in producing the similarity measure (or the goodness-of-fit). That is, one feature's contribution to the similarity measure depends on other feature(s)' contribution(s) to that measure.[21] For example, a similarity interaction is shown by the verification of salinity in the Bay model, in which the measurement of salinity (as a measurement of one feature's contribution to the similarity value from Weisberg's perspective) depended on the measurement of other features in the way that other features must be kept constant. For example, when verifying tidal current and salinity, "it was essential to obtain all data simultaneously for a given tide over at least one complete tidal cycle of 24.8 hours" (Army Corps 1963, 50), and "salinity phenomena in the model were in agreement with those of the prototype *for similar conditions of tide, ocean salinity, and fresh-water inflow*" (ibid., 54; my emphasis).

The two preceding quotations indicate two ways of keeping the other features constant when verifying one feature: (a) when obtaining data about a feature in one object, we keep the other features of this object constant (hence this guarantees that we obtain data about one feature under the same circumstance), and (b) when comparing the same feature of two objects, A and B, we keep the other features of A the same/similar as/to the other features of B (hence this guarantees that we compare one feature of two objects under the same circumstance). The first quotation instantiates the first way of keeping the other features constant, because obtaining "all data simultaneously" means that the obtaining of each feature's data was achieved against a similar circumstance. The second quotation instantiates the second way of keeping the other features constant, because when comparing salinity of the model with salinity of its target system the other features (e.g., tide, ocean salinity, and freshwater inflow) of the model and of its target system were kept similar.

This practice of keeping the other variables constant when verifying one variable is justified by the common methodology deployed in scientific practice, for example, in developmental biology scientists often keep the environmental factors constant when investigating the causal effect of a gene (or a set of genes) on the development of a phenotypic trait (e.g., eye colors). If one were to design a series of experiments to investigate how a gene might affect the development of a phenotype without keeping the environmental factors constant (or without against a similar environment), then it would be highly dubious whether the gene is really the difference-maker.

The causal interaction (CI for short) and the similarity interaction (SI for short) are two different things. To begin with, CI means that the causal effect of one factor depends on the causal effect of another factor, while SI means that one feature's contribution to the similarity measure (or the goodness-of-fit) of a model depends on other feature's or features' contribution to that measure.[22] The interplay between tide and salinity in the Bay model described earlier is a good example of CI, because how salinity produces its outputs depends on how tide varies along an estuary. On the other hand, the interplay between the measurement of salinity and the measurement of the other features in the Bay system (e.g., tide) is a good example of SI, because, as we often do in practice, we must keep the other features constant when measuring salinity. A good way to understand what SI is to first understand what SI is not. Suppose there is a concrete model involving only two factors such as salinity and tide. If, for example, we can measure salinity's contribution to the similarity measure of the model independently, that is, we can measure salinity's contribution without considering tide (e.g., without keeping tide constant), then the similarity measure of the model with respect to its target system can be achieved by simply adding the similarity measure of salinity and tide together. In this case, there is no similarity interaction between salinity and tide in the model. Second, CI concerns specific factors within an object (e.g., in a model or a target system), whereas SI concerns two objects because similarity concerns the degree to which one object is similar to another object.[23] Thus, they are two different concepts that should be distinguished clearly. Third, there is no simple relationship between the two. For example, when computing the fit of a straight line $y = ax + b$ to a cloud of points, a and b will depend on each other to produce the best fit.[24] Yet, although there exists similarity interaction between a and b, there is no causal interaction between them. However, CI and SI sometimes do happen together; for instance, in the Bay model, CI among the features goes hand in hand with SI among the same features in such a way that each feature's contribution to the similarity measure of the model depends on other features' contributions to that measure.[25]

To sum up, we have seen that features of a model may interact with one another to produce outputs. This causal interaction among features has led us to notice another kind of interaction, namely, the similarity interaction, wherein one feature's contribution to a model's similarity measure (or the goodness-of-fit) may depend on the other features' contributions to that measure.

6.4.2 *The Model–World Relationship Is a Holistic Matter*

The last subsection has shown that the features in the Bay model are not independent, and, more importantly, they interact with one another in producing phenomena (or outputs).[26] This leads us to a further claim: the Bay model bears a holistic fit to its target system. More specifically, due to the similarity interaction among the features discussed earlier, the modelers compared the Bay model with its target system holistically when verifying the model. To show this, let us go back to the verification scenario.

At first blush, it seems the verification of the Bay model was achieved by independently verifying each individual feature, as the report showed.[27] That is, it seems that by verifying that each feature in the model fits its counterpart in the target, the modelers made the judgment that the model fits the target system. Underlying this seemingly plausible reasoning, however, there remains a problem of why the modelers were allowed to confirm the verification of the model by means of only verifying several individual features. Or, to put it slightly differently, in terms of what does the fit of the features guarantee the judgment about the fit of the model to the target? I believe that it is more than the fit of those individual features themselves that makes sense of the reasoning that the model fits the target.

Looking more closely at the verification procedure, we obtain an answer to our question (of why the modelers were allowed to verify the model by only verifying several individual features). Namely, the modelers did not really verify each feature independently. Rather, they routinely kept the other features fixed when verifying one feature, and this was based on the recognition that there might be similarity interaction among the features (i.e., one feature's contribution to the similarity measure of the model might depend on the other features' contributions to that measure). Therefore, the actual scenario is not that the modelers verified the model by verifying each individual feature independently but rather that the verification of each individual feature involved the other features.

Not surprisingly, the scenario portrayed earlier dovetails with the actual verification procedure. To show this, it is worth quoting the sentences again: "Among the problems to be considered were ... the tides, currents and salinity of the Bay as they affect other problems.... *None of these problems can be studied separately, for each affects the others*" (Huggins and Schultz 1973, 12; my emphasis). It is clear that the modelers were well aware that, although each feature can be investigated separately, the contribution of each feature to the similarity measure of the model cannot be considered independently. Given the fact that the features might have similarity interaction, it is thus important to keep the other features constant when verifying one feature, a practice that has been shown in the verification of salinity: "salinity phenomena in the model were in agreement with those of the prototype *for similar conditions of tide, ocean salinity, and fresh-water inflow*" (Army Corps 1963, 54; my emphasis).

With this discussion in mind, it is not difficult to find the same salient characteristics as those found in the hypothesized causal structure about leaf gas-exchange. As in the MLE case, where we cannot compare two structures directly, in the case of testing the concrete object, we also cannot compare two structures directly. For this reason, the modelers collected data about each individual feature of the object (e.g., tide, salinity, and velocity), just as the modelers collected data about each variable in the hypothesized causal structure. Yet collecting data about each individual feature of the Bay model does not mean that the modelers did not know that the features might interact, nor does it mean that the interactions between the features were negligible. Rather, they simply took as a background assumption the fact that the features may interact, and the practice of keeping the other features constant when testing each individual feature has documented this. Thus, the first salient characteristic of testing the concrete object is that the modelers treated their model holistically. The second salient characteristic, which has been implied by the first, is that the modelers knew that the effect (i.e., the output) of each feature detected was not produced by that feature alone; rather, how one feature could produce effects must be constrained by how the other features behaved. In other words, the features interacted with one another in producing certain outputs. This CI has also led us to notice another kind of interaction in the Bay model, that is, an SI, wherein the features interacted with one another in producing the similarity measure of the model. In other words, the practice of testing a concrete object also satisfies the criterion for a holistic fit (see Section 6.1).

In conclusion, we have seen that the philosophical implications obtained by using MLE to test the hypothesized causal structure apply in the concrete model case, even though the testing of the concrete object did not involve MLE or something similar (note that the MLE method was developed in the late 1960s and the 1970s by Jöreskog [1967, 1969, 1970a, 1970b] and Keesling [1973], long after the testing of the Bay model in the 1950s). Given this, the lesson we achieved in the previous subsections can be further generalized: for many scientific models (biological or nonbiological), insofar as the model–world relationship is concerned we consider two structures holistically. And the reason we treat the model–world relationship in such a way is simply that we know (or at least we assume) that there must be such and such a holistic structure over there in the real world.

6.5 What Is a Structure?

Given that the term *structure* is extensively employed in the holistic view, one may wonder what it consists of. Moreover, the problem can become even more serious, if one rightly points out that my notion of structure seems to bear close affinity to the notion employed in the semantic view. Proponents of this view are talking about isomorphic, partial isomorphic, homomorphic relationships (or the like) between models and their target systems, and these relationships are also interpreted as structural relationships. That is, they are also relationships between two *structures* as my account suggests. So what do I mean when invoking the term *structure*?

To answer this question, my strategy is to first briefly outline the semantic view of models and then argue, by suggesting a deflationary approach to model structures, that my approach is more inclusive than the semantic view because the latter can be treated as a special case of my approach.

6.5.1 The Semantic View of Models

The semantic view was advocated as an alternative to the syntactic approach to scientific theories, suggesting that we focus on the semantics of theories rather than on the syntax of theories (Suppes 1960, 1962; Sneed 1971; Suppe 1977, 1989; Stegmüller 1976; van Fraassen 1980; Thompson 1983; Lloyd 1994; etc.). Instead of viewing scientific theories as deductive systems wherein axioms and theorems expressed in the first-order language bear deductive relationships, the semantic view thinks of a scientific theory as consisting of a family of models. The relationship between a theory and its models is that of *satisfaction*—namely, models make the theory expressed as a set of postulates, theorems, or axioms true (Downes 1992, 144; Odenbaugh 2008, 510).

With respect to the conception of *models*, the semantic view has two different answers: first, models are trajectories in state spaces,[28] and second, "models are set-theoretic structures defined by set-theoretic predicates"

(Odenbaugh 2008, 511).[29] Regarding the model–world relationship, the semantic view holds that it is isomorphism (or some weakened versions). Roughly speaking, isomorphism is a relation between mathematical structures, wherein there is a one-to-one function that maps each element of one structure onto each element of another structure while preserving the relations defined in each structure (Downes 1992, 147; Suárez 2003, 228).

As the semantic view developed, a number of authors questioned its plausibility in regard to understanding modeling practice. In particular, many authors point out that what the semantic view means by models is far removed from what scientists mean by models in the scientific context (e.g., Griesemer 1990; Downes 1992; Giere 1988; Cartwright 1999; Morrison 1999; Suárez 2003, 2004, 2016; Callender and Cohen 2006; etc.). Moreover, it is argued that the semantic view cannot accommodate the widely used idealizations in scientific modeling, which render any one-to-one mapping account a nonstarter (e.g., Frigg 2006; Odenbaugh 2008; Weisberg 2013). Facing these criticisms, some authors attempt to rescue the view by employing certain weakened versions of isomorphism, for example, partial isomorphism[30] and homomorphism (Lloyd 1994; Miller and Page 2007). Yet, as many point out, these weakened versions fare no better than their predecessor (e.g., Pincock 2005; Frigg 2006; Odenbaugh 2008; Suárez and Cartwright 2008; Weisberg 2013).

However, evaluating the plausibility (or implausibility) of the semantic view is surely beyond the scope of this chapter. Instead, in what follows, I suggest a deflationary approach toward model structures. The motivation behind proposing such a deflationary approach is to remain neutral in the debate about the semantic view, for, on the one hand, if the semantic view turns out to be true, then it can be treated as a special case of my approach, and on the other hand, if it turns out to be untrue, then my approach can be advanced as an alternative.

6.5.2 A Deflationary Conception of Model Structures

In this section, I suggest a *deflationist* view of model structures, a view also advocated by many other authors (e.g., Downes 1992; Suárez 2004, 2015, 2016; Godfrey-Smith 2006; Odenbaugh 2014; etc.). Largely following Downes's (1992) deflationary conception of models and modeling that says that "model construction is an important part of scientific theorizing" (151), I claim that model structures, of various kinds (e.g., concrete physical structures, equations, graphs, pictures, abstract structures in logic and metamathematics, etc.), are important *inferential tools* in modeling practice.[31]

Nevertheless, to be inferential tools, models must have one key feature: they are, or at least can be described as, *dependence relationships*.[32] Hence, my deflationary account comes down to the claim that models are important inferential tools that are, or at least can be described as, dependence relationships.

The dependence relationship can be causal if what are under consideration are causal models, such as concrete physical objects (e.g., the San Francisco Bay model), or mathematically described causal models (e.g., the leaf gas-exchange model). The dependence relationship can also be noncausal if what are under consideration are noncausal models, such as mathematical, optimality, minimal or equilibrium models (Rice 2015; also see Sober 1983; Batterman 2002a, 2002b; Bokulich 2008, 2011, 2012; Batterman and Rice 2014). As an example, consider Sober's equilibrium model in which a law of coexistence is invoked: a population with a certain required structure would *simultaneously* instantiate the Fisherian relationship (i.e., the 1:1 sex ratio) (1983, 204). Ohm's law,[33] the law for a simple pendulum[34] and the ideal gas law[35] are oft-cited cases of laws of coexistence, where the corresponding dependence relationships are noncausal (for a discussion of laws of coexistence see Hempel 1965, 347–354; Byerly 1990; Kistler 2013; etc.).

Moreover, the dependence relationship can also be relations in logic and meta-mathematics if what are in question are purely mathematical relationships. Unlike the noncausal dependence relationship, which invokes physical quantities (e.g., temperature or pressure), the dependence relationship in logic and meta-mathematics only invokes axioms or theorems within mathematics, although the dependence relationship might be applied to physical systems. As an illustration, consider briefly Baron, Colyvan, and Ripley's example involving intervening on mathematical truths and its implications for physical systems. In North America,

> [t]hree species of cicada of the genus *Magicicada* share the same unusual life-cycle. In each species the nymphal stage remains in the soil for a long period, then the adult cicada emerges after either 13 years or 17 years depending on the geographical area.
>
> (Baker 2005, 229)

One explanation for this unusual phenomenon has been provided by Goles et al. (2001), who

> hypothesize a period in the evolutionary past of *Magicicada* when it was attacked by predators that were themselves periodic, with lower cycle periods. Clearly it is advantageous—other things being equal—for the cicada species to intersect as rarely as possible with such predators

and "the frequency of intersection is minimized when the cicada's period is prime" (Goles et al. 2001; cf. Baker 2005, 230–231). According to Baron et al. (2017), there is a mathematical dependence relationship underlying this phenomenon, that is, being a prime number and maximizing the least common multiple of the prime number and any other numbers (being prime or not). Correspondingly, the relevant dependence relationship in the physical world involves the relationship between, for example, having a prime number life cycle and minimizing the frequency of encountering predators.

One may worry that my notion of structures is so permissive that it contains almost everything, including the "pseudo" dependence relationship that "had the reading of a barometer *B* been changed to such and such, there would have been a storm *S*". However, the notion is not so permissive as such; there exists a threshold over which genuine and spurious dependence relationships are delineated. That is, a genuine dependence relationship is one that enables one to make proper inferences; it must serve as an inferential tool in modeling practice. In particular, to use Woodward's framework and extend it to noncausal cases,[36] we may say that genuine dependence relationships (causal or noncausal) are those that allow one to answer what-if-things-had-been-different questions ("w-question"): they tell us how certain variables would change if other variables were to be changed (Woodward 1997, 2003, 2010; also see Woodward and Hitchcock 2003). The ideal gas law is a (noncausal) genuine dependence relationship in point, which describes, for example, how the temperature would change if we were to keep the volume constant and intervene on the pressure. By contrast, the pseudo dependence relationship between the reading of a barometer and a storm should be ruled out because it offers no help in answering w-questions—that is, changing the values of the antecedent variable (different reading of the barometer) does not tell us how the values of the consequent variable (storm/non-storm or different strength of the storm) would change.

In the causal dependence relationships case, the corresponding w-questions take the form of counterfactuals, such as "Had the specific leaf mass (SLM) in the leaf gas-exchange model taken a value other than what it really takes, then what would have the resultant internal CO_2 concentration been?" In the noncausal dependence relationships case, the corresponding w-questions take the same form as those causal dependence relationships, although the dependence relationships are noncausal.[37] For example, with respect to the ideal gas law, when holding fixed the temperature, the w-question is that "had the pressure taken such-and-such a value the volume would have taken such-and-such a value". Finally, in the mathematical dependence relationships case, the corresponding w-questions take the form of *counterpossible* statements,[38] for example, "had 13 not been a prime number (e.g., $13 = 2 \times 6$), then the least common multiple of 13 and 12 would have been smaller than $12 \times 13 = 156$". Putting this dependence relationship back to the North American cicada species, since it entails that having a prime number life cycle is a good strategy to avoid predators (because the prime number maximizes the interval of years at which the cicada species intersect with the predators and thus minimizes the frequency of intersection), the corresponding counterfactual statement is that "[i]f in addition to 13 and 1, 13 had the additional factors 2 and 6, then the North American cicadas would not have 13-year life-cycles" (Baron et al. 2017, 7).

Given this characterization of model structures, one might worry what goods my approach can deliver. First, because my approach attempts to accommodate different kinds of model structures, it is consonant with the heterogeneous nature of modeling practice. In particular, it avoids the criticism that the models in the semantic view are different from the models in the scientific context,[39] because those models in the semantic view can be treated as a special kind of models within my framework. That is, a model can be a concrete physical structure, a mathematically formulated causal structure, a noncausal structure, an abstract structure in logic and meta-mathematics, and so on, as long as it involves the dependence relationship that helps in answering w-questions. Second, the more challenging problem for the semantic view, that is, idealizations (Pincock 2005; Frigg 2006; Odenbaugh 2008; Weisberg 2013), does not arise in my view. This is because it follows from my view that model structures can either be *hyperaccurate structures*[40] that map to structures of their targets in an isomorphic manner or idealized (or distorted) structures that only *fit* parts or aspects of their targets (idealized parts, nonidealized parts or both) or even "nominal" structures used either in predictive modeling in which modelers are only interested in prediction accuracy[41] or in minimal modeling in which modelers are only interested in how to reproduce the overall pattern of behavior in different systems that have heterogeneous underlying details (Batterman 2002a, 2002b; Batterman and Rice 2014). Therefore, my view treats the notion of model structure as a continuum, ranging from the strict structures that can be mapped to their targets in a formal fashion (e.g., isomorphic, partial isomorphic or the like), through the idealized or distorted structures that merely preserve certain parts or aspects of their targets, to the nominal structures that might have nothing in common with the real structures of their targets.

Against this backdrop, one implication is that the holistic fit can also be construed deflationarily. That is, it can be strict isomorphism (or some weakened version) in very limited cases if what is under consideration is the relationship between a hyperaccurate model and its target system; it can be just a fitting between two outputs or two overall patterns if what is under consideration is a predictive or minimal model; or it can be something in between, which is neither a formal mapping nor a simple fitting of outputs or patterns. In each case, the degree of the holistic fit, or the distance between two dependence relationships, can be measured via various methods depending on different kinds of modeling.

In sum, a deflationary conception of structures merely claims that model structures, of various kinds, are important inferential tools in modeling practice. They can be concrete physical structures, equations, graphs, pictures or abstract structures in logic and meta-mathematics so long as dependence relationships are involved. It follows that the relationship between the model and its target, that is, holistic fit, can be recast through the deflationary lens.

6.5.3 Organization and Features

As mentioned in Section 4.2, one may point out that organization itself could be a feature, so a drawing of Tom's face capturing accurately not only his nose, mouth, and eyes but also their organization can be a good model of Tom's face. This point can be motivated as a response to my criticisms of Weisberg's view. Of course, a holistic account agrees that organization could be a feature but disagrees with the particular way in which organization is treated in Weisberg's similarity measure. Intuitively, we may say that a drawing of one person's face is a good model if it has the right features, such as a nose, a mouth, eyes and the organization of all these. So it seems that if one gets each individual feature right, then they get the whole model right. That is, it seems that features *additively* contribute to the goodness of the model.

This intuitive way of understanding scientific modeling, however, obscures the fact that features may interact in producing the fit of a model. To reiterate this point and to draw a connection to our current discussion, consider another ordinary example.[42] Suppose Anne's face is an ideal one which scientists want to model. Anne has an ideal nose, which is straight, in contrast to a nonideal nose, which might be bumped or concave. She also has an ideal nostril, which is round, in contrast to a nonideal one, which might be triangular or square. Scientist *A* draws a face for Anne that has a round nostril and a concave nose, while scientist *B* draws a face that has a triangular nostril and a bumped nose. Drawing *A* has an ideal feature (the round nostril), but neither feature of drawing *B* is ideal. Now we ask which drawing better fits Anne's face. It is likely that, due to the taste of our contemporaries (as part of the criteria when testing the model), we will say that *B* is better because the combination of a triangular nostril and a bumped nose produces a slightly less ugly image than that produced by the combination of a round nostril and a concave nose, although a round nostril itself is ideal. Hence we see a case wherein the nostril and nose interact to produce the fit of a model to a target.

This discussion leads to a more general question: What are features? In Weisberg's account, a model can *more or less* fit a target, but features are either shared or not. Yet, as Wendy Parker (2015) points out, "relevant similarities often seem to occur at the level of individual features, not just at the level of the model" (273). This is because features themselves can be objects such that they more or less fit each other.[43] Weisberg may argue that this problem can be fixed by the assumption that a feature can be redescribed as a set of sub-features, so the similarity between two features can be measured as the result of the similarity between their sub-features. However, I do not think this treatment is promising, for the similarity between sub-features may also be a matter of degree such that it should be measured as the result of the similarity between their sub-sub-features, between their sub-sub-sub-features, and so on. This regress is vicious because measuring the similarity between two features cannot be achieved without first measuring the similarity between their sub-features and measuring the similarity between their sub-features can also not be achieved without measuring the similarity between their sub-sub-features, and so on and so forth.

In contrast, a holistic account does not encounter this problem. If a feature is an object, then it can be simply treated as a structure. So the relationship between a feature in a model and a feature in a target also consists in their holistic fit. Take a minimal model, for instance. This example comes from Batterman (2002a). Most minimal models primarily attempt to represent repeatable patterns of behavior that are "largely insensitive to the underlying microscopic details" (Batterman 2002a, 27). Suppose we notice the buckling behavior of struts and find a formula to describe it (the formula is called Euler's formula):

$$P_c = \pi^2 \frac{EI}{L^2}, \tag{6.12}$$

where P_c is the critical buckling load, E is Young's modulus, I the second moment of the strut's cross-sectional area, and L the length of the strut (ibid., 27). The pattern of behavior in question is universal and insensitive to its underlying details: "The functional form—that the buckling load P_c is proportional to I/L^2—is universal. System specific details—that is, details dependent upon the microstructural make-up of the individual systems—are absorbed into the phenomenological parameter E, which will be different for struts composed of different materials" (ibid., 27).

As a first step, we have to make sure what the feature under consideration is. Let us suppose that the repeatable pattern of behavior, as expressed in Equation 6.6.12, is the only feature under concern. Then it seems that a proponent of Weisberg's view has nothing to say about this feature except that the model shares the feature with its target system or not. And, obviously, sharing or not is an all-or-nothing condition. By contrast, this situation poses no threat to the holistic approach, for even if the pattern of behavior is the only feature under concern, we are still able to consider the degree to which the model has this feature as its target does. More specifically, if we treat the pattern as an overall outcome of interactions between various variables (e.g., I and L), then the next step is to delve into the way that these variables interact to produce the outcome. To achieve this, we may first measure each variable separately and then calculate the covariances between each pair of variables. After this, we may appeal to certain testing methods to minimize the discrepancy between the expected covariance pattern (given derived data) and observed covariance pattern (given observed data) and then calculate the probability of having observed the minimum difference between the two patterns. As such, we are able to measure the degree to which this single feature in the model resembles its counterpart in the target system, and obviously this is no longer an all-or-nothing issue.

In short, a holistic approach to models fares better than its rivals (e.g., Weisberg's similarity view) because it can not only treat an overall system as a structure but also treat the components (or elements if the relevant relationship is not compositional) of the system as structures.

6.6 Conclusion

This chapter has proposed a holistic approach to models by way of looking closely at scientific practice. This approach views the model–world relationship as a holistic fit wherein a holistic fit stands for the degree to which the structure of the model resembles its counterpart in its target system. More precisely, for a good model, the model–world relationship constitutes a holistic fit when (a) it involves the situation in which the modelers take the model (and its target system) as an interconnected whole, wherein the components of the whole interact with one another in producing certain outputs (or patterns, phenomena, etc.), and (b) the goodness-of-fit measure of the model with respect to its target system is achieved via comparing the outputs (or patterns, phenomena, etc.) produced by these interacting components. Moreover, this chapter has shown that a deflationary conception of model structure can not only avoid thorny problems the semantic view faces, but also cast light on broader modeling practice. Ultimately, the deflationary view regards structures as dependence relationships, causal or noncausal. With this construal of model structure, the holistic fit between a model and its target system can be approached through the deflationary lens.

With this holistic approach, the question of why biological models can be explanatory can be answered through a new perspective: they are explanatory because they are in essence dependence relationships that can be exploited to answer Woodward's what-if-things-had-been-different questions. In particular, because they bear a holistic relationship to their target systems, the implications about the dependence relationships in the model can ramify across to their counterparts in the target system. That is, what we find in the model is expected to be found in the model's target system as well (and vice versa). These ideas are fully explored in the next chapter when a holistic account of model explanation is developed.

Notes

1 Depending on different modeling goals, "a good model" can mean a number of different things, for example, it can refer to a model with a highly desired prediction accuracy if what interests scientists is prediction, refer to a model

that has a highly realistic representation of the underlying mechanism of a phenomenon if what scientists are interested in is explanation, to a model with high generality if scientists are concerned with the explanatory scope of the model, and so on. Hence, in what follows, when I say "a good model", I mean that the model satisfies the given goal of a modeling practice.

2 For some predictive modeling practices, modelers may not be interested in the underlying structure of the target system; rather, the prediction accuracy of the model may be the only concern. For some minimal modeling, the overall pattern of behavior manifested by target systems is the only concern, which may be partially independent of its underlying causal details. Section 6.5 says more about these points.

3 Interpreted in mathematical parlance, measuring the degree to which one structure holistically fits another structure sometimes comes down to measuring the difference between two patterns of covariances (or correlations) among variables—one is the *observed* pattern, and another is the *predicted* pattern of covariances. As discussed in the last chapter using the leaf gas-exchange model, the essence of testing the goodness-of-fit of the model consists in comparing the predicted and observed covariance matrices.

4 I thank Wendy Parker for drawing my attention to the problem concerning the relationship between model evaluation and model adequacy.

5 For example, in the following cases, MLE is preferable to other methods: when the model includes latent variables, when the model must rely on observed variables that contain measurement errors, and so on (Shipley 2002, 71). Latent variables are those that cannot be directly observed and measured (ibid., 71).

6 Although the testing procedures are different (we will see in this and the next sections), the underlying idea of LSE is based on MLE. Moreover, when we assume that the measurement errors are independent and normally distributed with constant standard deviation, LSE is equivalent to MLE (for a discussion of the relationship between LSE and MLE, see Press et al. 1992, Chapter 15).

7 For a discussion of nonlinear LSE, see Pollard and Radchenko (2006).

8 I thank Arnaud Pocheville for bringing this to my attention.

9 This point is called an *outlier* in statistics, representing an observation point that is far from the bulk of data. This may be due to improperly calibrated equipment, recording and transcription errors, or equipment malfunctions, among others. (Rice 2006, 393–394).

10 $N(0,\sigma)$ means "a normally distributed random variable with a population mean of zero and a population standard deviation of σ". For details see Appendix 1, Note 1.

11 For the details of the probability density function, see Shipley (2002, 110–114) or Rice (2006, 329–334).

12 For details of the likelihood function, see Shipley (2002, 113) or Greene (2012, 549–550).

13 For details of this function, see Shipley (2002, 113–115).

14 For the calculation of degrees of freedom, see Section 5.4.

15 The testing process involves using the maximum likelihood chi-square statistic, which will asymptotically follow a central chi-square distribution with the degrees of freedom. For details of this statistic see Shipley (2002, 114–115).

16 For the meaning of significance level, see Section 5.4.

17 This metaphor comes from Shipley (2002, 1–6).

18 "Equivalent models are those that provide the same sets of statistical fit indexes (e.g., chi-square and p values) as a hypothesized model but may imply

very different substantive interpretations of the data" (Stelzl 1986; cf. Raykov and Marcoulides 2001, 142). For a discussion of equivalent models see Stelzl (1986), Lee and Hershberger (1990), MacCallum et al. (1993), Raykov and Marcoulides (2001), and Bekker et al. (2014).

19 Following Michael Weisberg, I think features refer to two different categories of things, with one category standing for properties and patterns and another standing for the underlying mechanisms that generate the properties and patterns in the first category. The properties and patterns in the first category can be conceptualized as states and the underlying mechanisms as transition rules (for details of this distinction, see Weisberg 2013, 145–146). Since the verification of the San Francisco Bay model in the 1950s concentrated on properties such as salinity, velocities of currents and tide, not on the underlying mechanisms that generated these properties, the term *feature* in this section narrowly denotes the first category.

20 I thank Arnaud Pocheville for suggesting this important concept to me and for giving me the idea about what the similarity interaction is and in what sense it differs from the causal interaction. Wendy Parker (2015, 274) has a similar idea to "similarity interaction" when evaluating Michael Weisberg's similarity measure, although she does not develop this idea.

21 This point can be best illustrated with the curve fitting example: when computing the fit of a straight line $y = ax + b$ to a cloud of points, a and b will depend on each other to produce the best fit. I thank Arnaud Pocheville for giving me this example. This point has been shown in Section 6.2 with LSE.

22 I thank Arnaud Pocheville for giving me these precise definitions.

23 I thank Arnaud Pocheville for illustrating this difference.

24 I thank Arnaud Pocheville for giving me this example.

25 The relationship between CI and SI may be more complicated than what has been presented in this section; for example, there might be cases in which there is CI while there is no SI. However, noticing that CI and SI sometimes do happen together is sufficient for my current purposes (i.e., showing that in the Bay model, there exist both CI and SI), and addressing the more complicated issue must be left to another occasion. I thank Arnaud Pocheville and Paul Griffiths for letting me pay attention to the complicated relationship between causal and similarity interactions.

26 In what follows, by *features*, I mean these three features: salinity, velocities of currents and tide.

27 See Section 4.2 for the verification report.

28 "A state space consists in the set of all the possible values of the variables. If each variable is given a determinate value, then there is a particular point in that space which is the *state* of the system. We can represent this state of the system at t as a vector X_t and its dynamics as a sequence of such states or vectors. Likewise, there are also parameters that mediate the relationships between variables" (Odenbaugh 2008, 511).

29 "For example, a model of the kinetic theory of gas … can be construed as a structure that is in the extension of the set-theoretic predicate $<P, T, s, m, f, g>$" (Odenbaugh 2008, 511).

30 See van Fraassen (1980), Bueno (1997), French (1997, 2003), French and Ladyman (1998), Bueno et al. (2002, 2012), da Costa and French (2003), Bueno and French (2012), etc.

31 Mauricio Suárez (2004, 2015, 2016) and Bueno and Colyvan (2011, also see Bueno and French 2012; Baron et al. 2017) also develop their inferential accounts of models. The former emphasizes the use-based aspect, while the latter features the partial isomorphism approach in which it is the partial isomorphic relationship between the model and its target that guarantees the

inference from the former to the latter. It is worth mentioning that Suárez's account is also deflationary, in the sense that it construes representation as *means*—rather than *constituents*—of inference-making, implying that representation can be fulfilled by any means of inference (such as induction, deduction, abduction, etc.) without committing to any constitutive relationship (such as isomorphism, partial isomorphism, homomorphism, etc.) between the model and its target.

32 Christopher Pincock proposes a similar idea ("objective dependence relations") when discussing how noncausal explanations work. According to him, in addition to causal dependence relations there are abstract dependence relations that can also be used to do explanatory work (Pincock 2015, 878). The next chapter shows how the understanding of structures as dependence relationships can be employed to develop a holistic account of model explanation.

33 That is, $V = IR$, voltage is the product of current and resistance.

34 That is, $T = 2\pi\sqrt{L / g}$, where T refers to the period, L to the length and g to the gravitation constant.

35 That is, $PV = nRT$, where P denotes the pressure, V to the volume, R the gas constant, and T the temperature.

36 Woodward (2003) takes an open attitude toward extending his framework to noncausal cases, as he states that "when a theory or derivation answers a what-if-things-had-been different question but we cannot interpret this as an answer to a question about what would happen under an intervention, we may have a non-causal explanation of some sort" (221). Recently, there are a number of attempts to extend Woodward's framework to noncausal explanations, for example, Frisch (1998), Bokulich (2008, 2011, 2012), Rice (2012, 2015), Kistler (2013), Saatsi and Pexton (2013), Pexton (2014), Pincock (2018), and Reutlinger (2016, 2017), among others.

37 For a discussion of how non-causal explanations can be explanatory, see Steiner (1978), Jackson and Pettit (1990), Colyvan (2001, 2002, 2012), Baker (2005, 2009), Pincock (2007, 2011, 2015), Lyon and Colyvan (2008), Lyon (2012), Lange (2013), Baron et al. (2017), and others. For a discussion of how noncausal dependence relationships (i.e., noncausal models) can be employed to explain, see Sober (1983), Batterman (2002a, 2002b, 2010), Bokulich (2008, 2011, 2012), Rice (2012, 2015), and others.

38 It is called counterpossible because it contradicts with the mathematical theories of prime numbers in which 13 is a prime number. For details, see Baron et al. (2017).

39 For the criticism, see Giere (1988), Griesemer (1990), Downes (1992), Suárez and Cartwright (2008), Weisberg (2013), and others.

40 Hyperaccurate models are those that represent almost everything of their target systems. See Weisberg (2013, 150–151) for the relevant discussion.

41 In this predictive model case, the dependence relationship consists solely in the relationship between inputs and outputs.

42 I thank Arnaud Pocheville for giving me this nice example.

43 I thank Arnaud Pocheville for bringing this to my attention.

7 How Biological Models Are Explanatory

7.1 Introduction

The deductive-nomological (DN) model of scientific explanation claims that a generalization can be explanatory because it shows how the explanandum can be deduced from the explanans given certain initial/background conditions and at least one law of nature, that is, it shows how "the explanandum was to be expected" (Hempel and Oppenheim 1948). A classic objection to this view is the case in which a flagpole casts a shadow on the ground, and the sun's position and the shadow's length seem to explain the height of the flagpole, according to the DN model of explanation. This is because the explanation involves a geometric law, from which, together with the sun's position and the shadow's length, the height of the flagpole can be deduced. But, as discussed in Chapter 3, many philosophers point out that the DN model is an inadequate account of scientific explanation because it fails to capture this asymmetric feature of scientific explanation (e.g., Salmon 1989).

One way to solve this problem, suggested by James Woodward (1997, 2000, 2001, 2003, 2010), is to say that explanatory generalizations (or descriptions, equations, models, etc.) should be *change-relating* in the sense that they can be exploited to answer a range of "what-if-things-had-been-different questions" ("w-question"). That is, they can tell us information about how changes in variables that figure in the explanans, typically under intervention, can be systematically associated with changes in variables that figure in the explanandum, that is, provide us with *patterns of counterfactual dependence*. They should also be *invariant* in the sense that, although explanatory generalizations can in principle be intervened upon to answer w-questions, the relationships themselves involved in these generalizations can somehow be kept unchanged.

The interventionist approach seems very promising, and, on the basis of it, a structural account of model explanation recently has been built (Bokulich 2008, 2011, 2012).[1] On this structural account, a model "explains the explanandum by showing how the elements of the model

DOI: 10.4324/9781003148029-7

correctly capture the pattern of counterfactual dependence of the target system" (Bokulich 2011, 39). I think this account is largely on the right track, but it is more of a theoretical sketch than a full-blown view of model explanation. This is because, although this account rightly captures the intuition that the model is explanatory because it captures the pattern of counterfactual dependence of its target system, it does not flesh out how the model can do so.

To fully flesh out that idea, the holistic account developed in the previous chapter comes into play. The holistic account, broadly construed, is also a structural account of models. According to the holistic account, a model is just a structure, and a structure is a set of dependence relationships, causal or noncausal. The holistic account cannot only capture the intuition that a model can be explanatory because it can not only "capture the pattern of counterfactual dependence of the target system" (ibid., 39) but also show the specific way whereby the model captures the pattern of counterfactual dependence of its target. That is, it is the holistic fit between the model and its target that guarantees the (hypothetical) inference that the pattern of counterfactual dependence in the model may reappear in its target system (or vice versa).

In what follows, Section 7.2 first briefly outlines Woodward's interventionist account of explanation, followed by Section 7.3 describing Bokulich's structural account of model explanation. Then, in Section 7.4, based on Woodward's account and Bokulich's account, a full-fledged holistic account of model explanation is developed. To show how the holistic account really works, an agent-based simulation model drawn from biology is also examined in Section 7.4. Finally, Section 7.5 considers the question of whether this holistic account of model explanation can deliver any extra goods that its competitors cannot.

7.2 Woodward's Interventionist Account

As its name suggests, Woodward's (2000) account binds together explanation and intervention (also called manipulation or control): "explanatory relationships are relationships that *in principle* can be used for manipulation and control in the sense that they tell us how certain (explanandum) variables would change if other (explanans) variables were to be changed or manipulated" (198; author's emphasis).

Based on this understanding of explanation,

> It follows that whether or not a generalization can be used to explain has to do with whether it is *invariant* rather than with whether it is lawful. A generalization is invariant if (i) it is ... change-relating and (ii) it is stable or robust in the sense that it would continue to hold under a special sort of change called an *intervention*.
>
> (ibid., 198; author's emphasis)

"Change-relating" means that the antecedent variables in the explanation can *in principle* be intervened on in a way that systematically changes variables in the consequence. Some interventions are in principle possible because, although it may be impossible at present due to certain reasons, it is conceivable that had the antecedent been different in such and such a way, then the consequence would have been different in such and such a way. Suppose, for example, heavy rain in Sydney caused a flood in March 1980. We may not be able to intervene on the heavy rain, nor are we able to intervene on something that happened in the past. But it is still conceivable that, had it never rained, Sydney would not have been flooded. In contrast, some events or generalizations associated with fundamental physical theories are not change-relating, though they are explanatory.[2] For example, it seems unclear how we could intervene on the fundamental physical law: nothing travels faster than the speed of the light.[3] Another kind of generalization that is non-change-relating is the accidental generalization, which I say more about in the following.

One may worry about the notion of intervention, for, taken literally, it sounds anthropocentric. But it is not necessarily anthropocentric, because apart from human beings' activities, nature itself sometimes "intervenes" on objects in various ways, such as raining, flooding, thundering, and so on. More precisely, Woodward says that

> [a]n intervention is an exogenous causal process that changes some variable of interest *X* in such a way that any change in some second variable *Y* occurs entirely as the result of the change in *X*.... We may think of explanation as having to do not with subsumption under laws but rather with the exhibition of *patterns of counterfactual dependence of a special sort*, involving active counterfactuals—counterfactuals the antecedent of which are made true by interventions.
>
> (ibid., 199; my emphasis)

To see the connection between explanation and intervention, let us return to the flagpole case. Nature itself can intervene on the Sun's position and the length of the flagpole's shadow in various ways depending on the time of the day or the season of the year. But obviously, there is no change-relating relationship between the flagpole's height, on the one hand, and the sun's position and the length of the flagpole's shadow on the other, because intervening on the latter leads to no change in the former: the flagpole's height remains the same. For this reason, it is argued that the relationship involved in this example cannot be used to explain the height of the flagpole by the sun's position and the length of the shadow, because it cannot answer questions about what the consequence would have been had the antecedent been changed to such and such; that is, it fails to answer w-questions. Consider another case involving a common cause.

Suppose that a fall in atmospheric pressure *A* causes both the reading of a barometer *B* and the onset of a storm *S*. Because there is always coincidence of *B* and *S*, there seems to be a counterfactual dependence relationship among them: had it been the case that *B*, then it would have been the case that *S*, or had it been the case that *S*, then it would have been the case that *B*. But this dependence relationship can easily prove to be "pseudo" using Woodward's criterion: it is not change-relating because intervening on *B* does not result in associated changes in *S* and vice versa; that is, it cannot answer questions about what the consequence would have been had the antecedent been changed to such and such. Because of this, it is not an explanatory relationship.

Change-relating is the first criterion for a relationship or generalization to be explanatory;[4] another is *invariance*. That is, the relationship would continue to hold under interventions or changes. So an explanatory relationship should not only convey information about *patterns of counterfactual dependence* (i.e., can answer w-questions) under interventions or changes but also remain unchanged under these interventions or changes. A candidate in this case is Hooke's law:

$$F = -K_s x, \tag{7.1}$$

where F is the force needed to extend or compress a spring, K_s is the specific spring constant and x is the distance extended or compressed. Equation 7.1 conveys information about patterns of counterfactual dependence: if one were to change x by the quantity Δx, then F would change by the quantity $K_s \Delta x$. So the antecedent variable x is systematically counterfactually connected with the consequence variable F: supposing one were to intervene on x, one could answer a range of questions about how F would change. This relationship between F and x is also invariant, in the sense that given a certain range of changes of (or interventions on) x, the relationship itself still holds.[5] Yet, it is possible that there are points at which the relationship would break, for example, if one were to extend the spring so far that it cannot return back, and so on. In other words, the relationship holds over a certain range of changes but not others. Hence, the notion of invariance comes in different degrees, within which the relationship holds while outside of which it simply breaks.

Although invariance comes in degree, this does not mean that all relationships are on the same boat. Rather, there is a threshold between relationships that are invariant in various degrees and relationships that are not at all.[6] For noninvariant relationships, the flagpole example and the barometer reading example are two cases in point, and Clinton's pocket is another:

All the coins in Bill Clinton's pocket on January 8, 1999, are dimes.[7] (2)

This relationship is not invariant because, to use Woodward's (2000, 208) terminology, it does not describe a relationship hypothetically exploitable for the purpose of manipulation and control. That is, intervening on one variable that figures in the description (i.e., putting coins in Clinton's pocket) does not tell us how another variable (i.e., the fact about dimes) would change. For example, it is not difficult to imagine that had a penny been put into Clinton's pocket, it would not have been the case that all the coins in his pocket are dimes.

To summarize, a generalization is explanatory if it is both (a) change-relating and (b) invariant under interventions.

7.3 The Structural Account of Model Explanation

The idea that a model can be explanatory because it captures the pattern of counterfactual dependence of its target system at least can be traced back to Margaret Morrison's (1999) work on models, where she claims that "[t]he reason models are explanatory is that in representing these systems, they exhibit certain kinds of structural dependencies" (63). Yet Morrison does not develop this sound idea into a philosophical account of model explanation.

Partly due to the popularity of Woodward's interventionist view of scientific explanation, which is a version of the counterfactual view of explanation, the development of Morrison's idea has become possible very recently. Alisa Bokulich is one author who tries to bring Morrison's idea into a flesh-and-blood form based on Woodward's interventionist framework. Bokulich's (2011) basic idea is that a model can be explanatory because the model "explains the explanandum by showing how the elements of the model correctly capture the pattern of counterfactual dependence of the target system" (39).

For Bokulich (2008, 145), what makes an explanation an example of model explanation is that the explanans in question must make appeal to a scientific model that involves some degree of idealization. Given such, then a general account of model explanation must explain how a model can be genuinely explanatory and must be able to demarcate models that are explanatory from those that are not. Her answer has been mentioned, that is,

> [t]hat model explains the explanandum by showing how the elements of the model correctly capture the pattern of counterfactual dependence of the target system. More precisely, in order for a model *M* to explain a given phenomenon *P*, we require that the counterfactual structure of *M* be isomorphic in the relevant respects to the counterfactual structure of *P*. That is, the elements of the model can, in a very loose sense, be said to 'reproduce' the relevant features of the explanandum phenomenon. Furthermore, as the counterfactual

> condition implies, the model should also be able to give information about how the target system would behave, if the elements described in the model were changed in various ways.
>
> (Bokulich 2011, 39)

Here we see both Morrison and Woodward's influence on Bokulich. Interestingly, we can also see the influence of the semantic view on Bokulich from the presentation of her view, although it remains unclear if she really adopts this view. Finally, for Bokulich (2008), an adequate account must satisfy a further "justificatory step", "specifying what the domain of applicability of the model is, and showing that the phenomenon in the real world to be explained falls within that domain" (146). I return to the point about justificatory step in the next section.

To see how Bokulich's (2011) account works, let us consider her example: Niels Bohr's model of the hydrogen atom:

> According to Bohr's model, the electron can orbit the nucleus only in a discrete series of allowed classical trajectories known as stationary states. While in a stationary state the energy of the electron is constant, and the electron can only gain or lose energy by jumping from one stationary state to another. When such a transition or "quantum jump" occurs, a single photon of a given frequency is emitted (or absorbed). The frequency of the photon is given by the difference in energy of the two allowed orbits. The spectrum of hydrogen ... is built up out of the photons being emitted in these jumps between stationary states, where only those frequencies (or wavelengths) corresponding to allowed quantum jumps occur, and the intensity (or brightness) of a spectral line is given by the probability of that jump occurring. So, for example, the red spectral line $H\alpha$, is the result of the electron jumping from the $n = 3$ orbit to $n = 2$ orbit, and the green spectral line $H\beta$ is the result of transitions from the $n = 4$ to $n = 2$ orbit.
>
> (41)

Bokulich says her account of model explanation can cast light on why Bohr's model is genuinely explanatory. To begin with, the explanans makes an appeal to an idealized model, that is, Bohr's model; hence, it is clearly an example of model explanation. Moreover, the model "explains the explanandum by showing that the counterfactual structure of the model is isomorphic (in the relevant respects) to the counterfactual structure of the phenomenon" (Bokulich 2011, 43). This means that the model is able to answer a wide range of "w-questions" about its target system. For example, the model is able to answer questions such as "how the spectrum would change if the orbits were elliptical rather than circular, or how the spectral lines would change if the hydrogen atom were placed in an external electric field" (ibid., 43), and the like. Finally, there is a justificatory step that specifies the model's domain of applicability, and "the model is an adequate guide to that domain of phenomena" (ibid., 43). In particular, "modern semiclassical mechanics provides a top-down justificatory step showing that Bohr's model—despite failing as a literal description—is nonetheless a legitimate guide to quantum phenomena in certain domains" (ibid., 43).

There is one thing to note about Bokulich's account. As mentioned earlier, we can see the influence of the semantic view on Bokulich from the presentation of her view and noted that it is unclear if she really embraces this view. In several places, she indicates her awareness of the problems of the semantic view. For instance, she claims that "the counterfactual structure of the model is isomorphic (in the relevant respects) to the counterfactual structure of the phenomenon" (ibid., 43), and in a footnote, she adds that "[t]he notion of isomorphism is being used loosely here" (ibid., 43). Moreover, one author interprets her account of the model–world relationship as structural similarity (King 2016), which can be understood as less demanding than the isomorphism (or the partial isomorphism) required by the semantic view.

Therefore, a tentative interpretation of her view is that she does not assume the semantic view of the model–world relationship. However, even if this is true, we should still be very careful about the temptation to commit the semantic view of models and of the model–world relationship. As discussed in the previous chapter, the heterogeneous nature of models and modeling calls for a more comprehensive treatment (and for this reason a deflationary view was proposed there in order to cast light on as many different kinds of models as possible). After all, many models do not have such a formal relationship (i.e., isomorphism) with their targets. A case in point is the predictive model that only concerns prediction accuracy. As noted in the previous chapter, a predictive model may have nothing to do with the underlying structure of its target system, and the failure to share the same structure does not necessarily impair the model's ability to make precise predictions. Under these circumstances, a formal relationship such as isomorphism is not required. However, one may say that a predictive model still shares something with its target system; for example, the model produces an output that is similar to the output produced by its target system to a certain degree. That is, the model shares the output with its target system to a certain degree, or to put it another way, this case concerns the similarity of outputs. Yet, this is another sense of sharing (or similarity), one that must be distinguished from what the isomorphism or partial isomorphism approach requires (I come back to this point in Section 7.5).

In sum, although Bokulich's account correctly captures the intuition that a model can be explanatory because there exists some *link* between the counterfactual structures of the model and its target, it fails to provide a story of what that link ultimately consists in and thus fails to make sense of the fact that models can be explanatory. To fill in the gap, the next section will suggest an alternative view about how models can be explanatory.

7.4 A Holistic Account of Model Explanation

The previous chapter suggested that the model-world relationship is a holistic fit, wherein *holistic fit* refers to the degree to which one structure resembles another structure or refers to the distance between two structures (or outputs, patterns). The underlying idea is that if a model is a good model, then it should fit its target system holistically (or if the model is a good predictive model, then its prediction should be as accurate as required with respect to the output of its target system, or if the model is a good minimal model then it should produce the same pattern as detected in its target). The previous chapter also suggested that the holistic fit can take many different forms. That is, it can be strict isomorphism (or some weakened version) if what is under consideration is the relationship between a hyperaccurate model and its target system, it can be just a fitting between two outputs or patterns if what is under consideration is a predictive or minimal model or it can be something in between (due to idealization or distortion), which is neither a formal mapping nor a simple fitting of outputs or patterns. In either case, the structures—deflationarily understood as dependence relationships that serve as important inferential tools in modeling practice—of the model and its target bear a holistic relationship to one another.

Based on this holistic view of the model-world relationship, my central claim about model explanation (ME) boils down to the following statement:

> (ME) It is the holistic fit between the model and its target system that supports the *inference* from the model to the target, that is, supports the *hypothesis* that the counterfactual structure of the model may reappear in its target system.

The inference consists of two steps. First, since a model is a structure (as is a target system), that is, a dependence relationship, it follows that variables (or factors) in the model counterfactually depend on each other. More specifically, changes (or interventions) in explanans variables that figure in the model can be systematically associated with changes in explanandum variables that sometimes take the form of outputs of the model. The explanandum variables (or outputs, patterns), represented in the model, are supposed to describe (or reproduce) their counterparts in the world. As such, the model can be used to answer Woodward's w-questions about itself: we can ask how one variable in the model would change as a result of intervention on another variable in the

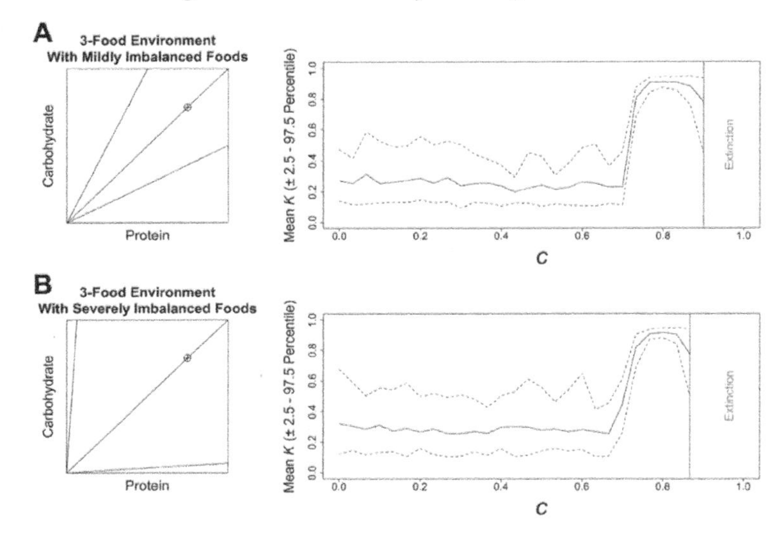

Figure 7.1 Results for the agent-based simulation model. Lines and crosshairs describe food rails and the intake target.

Sources: This figure comes from Senior et al. (2015, 7). Figure used with permission.[10]

model. Second, because of the holistic fit between the model and its target system, namely, the structure of the model resembles the structure of its target (to a certain degree), modelers believe that the counterfactual dependence relationships derived from the model may generalize to their counterparts in the target. In other words, a kind of inferential relationship has been hypothesized between the model and its target linking the two counterfactual structures.

In what follows, I show how this account works in a biological model. The model is drawn from Senior et al. (2015).[8] We may observe a common pattern in many arthropods (e.g., spiders, burying beetles, etc.), such as "the effects of contest competition and the number and composition of foods in the nutritional environment on the evolution of individual nutritional strategies" (ibid., 4–5; the pattern has been shown in Figure 7.1). Then we might attempt to understand "how intra-specific competition might affect the evolution of animals' nutritional strategies" (ibid., 4). According to Lihoreau et al. (2014),

> [a]n individual's nutritional strategy was governed by the fixed global parameter K, which we refer to here as 'nutritional latitude'. When eating a food that will not guide its nutritional state to the IT [Intake Target] an individual has some probability of leaving, which is both a function of the balance of nutrients in the food being consumed, and K. Here, a high K means an individual is likely to consume the same imbalanced food until reaching a point of nutritional compromise

> (at which point it then seeks an alternative). In contrast, a low K corresponds to a low probability that an individual will continue feeding on a food rail that will not guide its nutritional state directly to the IT.
>
> (Lihoreau et al. (2014); cf. Senior et al. 2015, 4)

The IT mentioned in the quotation refers to a coordinate or a region within the nutrient space, which denotes "the optimal amount and blend of nutrients that the animal requires over a specified period in its life" (Senior et al. 2015, 3). Now consider an agent-based simulation model employed to "explore how intra-specific competition might affect the evolution of animals' nutritional strategies" (ibid., 4).[9] The following is the model description:

> Each generation consists of 150 individuals that must attain a certain level of fitness (i.e. nutritional state) within a fixed number of model iterations for it to be considered fit enough to breed. Fitness-proportionate selection then operates among those individuals fit enough to breed, with proximity to the IT (optimal point of nutrient intake in the nutrient space) determining its fitness. We allowed K to evolve 1000 generations under varying levels of competition and in differing nutritional environment (i.e. different abundance and nutritional compositions of food). In doing so, we aimed to explore the effects of contest competition and the number and composition of foods in the nutritional environment on the evolution of individual nutritional strategies.
>
> (ibid., 4–5)

Suppose the population under consideration only feeds on three kinds of food, and we perform the model runs under varying intensities of competition, c, which is bounded at 0 and 1. The population mean nutritional latitude K obtained from each model run is also bounded at 0 and 1 (ibid., 5). Suppose, for simplicity, the environment contains one nutritionally balanced food (e.g., it contains the same amount of protein and carbohydrate) and two imbalanced but complementary foods (e.g., one might contain 40% protein and 60% carbohydrate while the other might contain 60% protein and 40% carbohydrate). For the latter two complementary foods, we can vary the extent of their nutritional imbalance to be either moderate or extreme (see Figure 7.1).

The results in Figure 7.1 are summarized as follows:

> In these environments when $c = 0$, K was stable at a range of values.... The high variance in stable values of K suggests that no one level of nutritional latitude is optimal where competition is weak, but most low levels are equally fit. In the face of increasing c, K was relatively stable up to a point. With mildly imbalanced foods at $c =$ 0.7, and with extremely imbalanced foods at $c = 0.67$, K increased

> sharply to above 0.91.... At very high c the population could not support itself as no individuals could fulfil the fitness requirements to be considered in breeding condition by the end of the simulation.
> (Senior et al. 2015, 5)

That is the outline of the agent-based model. Now let us go back to my claim that models can be used to answer counterfactual questions about themselves, that is, can answer how explanans variables that figure in the model can be systematically associated with changes in explanandum variables. In the agent-based model described earlier, there are three variables: competition for food (c), food composition (r), and nutritional latitude (K). The values of both c and K range continuously from 0 to 1 while r only takes two values in our model: mildly imbalanced foods (r_1) and severely imbalanced foods (r_2). First, consider the case where the food environment takes the value r_1. In r_1, for example, we may ask ourselves: if we were to change the value of c from c_1 to c_2, what would happen to the value of K. In particular, we may ask if we were to change the value of c from 0.1 to 0.2, what would happen to the value of K; if we were to change the value of c from 0.2 to 0.3, what would happen to the value of K; and so on. There are two points that may particularly interest us, that is, when $c = 0.7$ and $c = 0.9$, because these are points at which K changes drastically. The corresponding counterfactual questions are if we were to change the value of c from 0.6 to 0.7, what would happen to the value of K, and if we were to change the value of c from 0.8 to 0.9, what would happen to the value of K. Based on these counterfactuals and their corresponding answers in r_1, a systematic counterfactual dependence relationship between c and K has been built, which is summarized as follows:

$$c = 0, K(mean\ value) = 0.27$$

$$0 < c < 0.7, K = 0.271$$

$$c = 0.7, K = 0.91$$

$$0.7 < c < 0.9, K = 0.95$$

$$c \geq 0.9, K = 0$$

The similar situation holds for the case where the food environment takes the value r_2. To put it in Woodward's terminology, this means that changes (or interventions) in explanans variables can be systematically associated with changes in explanandum variables (or model outputs). Moreover, the established pattern of phenomena is *invariant* in the sense that it holds over certain range(s) of changes that (1) intervene on the initial and/or background conditions of the model and (2) intervene on the values of the explanans variables. For example, the pattern of phenomena does not only hold for colonies of social spiders and burying beetles but also hold for many other breeding vertebrates such as mongooses (*Mungos mungo*) and meerkats (*Suricata suricatta*; Senior et al. 2015, 2).

But that is just the first part of the story, because a model's explanatory power essentially relies on how the model is related to its target system.[11] That is, to be explanatory, a model should also support building inferential relationships that bridge the model and its target: for a good model, if the model behaves in such and such a way, then the modelers *hypothesize* that the target may also behave in such and such a way.[12] Here the holistic fit serves as the *inferential basis* bridging the model and its target. I call this kind of inferential relationship a *hypothetical relationship* (HR) or simply *hypothesis*, both because it features *hypothetical reasoning* in modeling practice,[13] and it directly relates to Giere's notion of *theoretical hypothesis* that, similarly to my own notion, links the model to its target in the real world. Giere (1988) says, "My preferred suggestion, then, is that we understand a theory as comprising two elements: (1) a population of models, and (2) various hypotheses linking those models with systems in the real world" (85). The first element is called a *theoretical definition* while the second a *theoretical hypothesis*.

By and large, hypothetical relationships are useful *heuristics* whereby scientists make inferences (i.e., explanations and predictions) about the world using the model. These HRs often take the following form:[14]

(**HR**):

i. If a model *M* holistically fits its target system *T* (i.e., if *M* is a good model),
ii. and, if *M* has such and such attributes, patterns, or mechanisms,
iii. then, *hypothetically*, *T* would also have such and such attributes, patterns, or mechanisms.

Note that the hypothetical relationship can run in many different ways: from the model to its target, or from the target to its model, or from both. Also, given that the building and adjusting of a model is typically an iterative process, the process of constructing the HR may also run back and forth between the model and its target system.

For HR, the fact that, given *M* is a good model, and *M* has such and such attributes, patterns, or mechanisms (attributes for short), can lead to the *prediction* that *T* may also have such and such attributes. Furthermore, exploring the way the model produces such and such attributes leads to the *explanation* of why *T* manifests such and such attributes. So part of the reason why a model can be explanatory (and predictive) is that, due to the hypothetical relationships that have been built between the model and the target, the modelers believe that the counterfactual structure of the model may be extrapolated to its target system. This results in the reasoning in modeling practice that if the model is a good model and if the model behaves in such and such a way, then, hypothetically, the target would also behave in such and such a way.

Let us take a close look at what guarantees the hypothetical move from the model to its target. We may ask what this kind of hypothetical relationship consists in. As said earlier, the counterfactuals concerning the model are rooted in the structure of the model itself. By assuming that a model is a good model, we are, in fact, assuming that the model resembles the structure of its target to a certain degree. That is, we are assuming that the model and its target have the same pattern of dependence relationships to a certain degree. This entails that the same interventions or changes on the variables of the model and the target will lead to the same changes in the outcomes of both the model and the target; that is, the outcomes we get from the model by changing certain variables should also be manifested in the target by changing the corresponding variables. In short, "twiddles in the model ramify across to the target system at issue"—this phrase comes from Baron et al. (2017), who express the same idea when discussing how mathematics can play an explanatory role in the physical sciences. The core idea is that, due to certain *morphism* relationships (e.g., partial isomorphism) between the mathematical and physical structures, twiddles to the mathematics ramify across to the physical system Baron et al. (2017). Although I agree with them that twiddles in the model ramify across to the target, I disagree that this is because of a sort of morphism relationship (in what follows I say more about this point).

So ultimately the hypothetical relationships that allow the inference from the model to its target consist in the *hypothesis* that the model structure resembles the structure of its target to a certain degree (and the previous chapter showed how to measure the hypothesis, i.e., the degree to which one structure resembles another). Note that the first kind of relationship (concerning the model itself) can be utilized as a foundation for building hypotheses about the second kind (concerning the model–world relationship, i.e., the HRs). Consider the agent-based model again. Suppose that the model is very good; that is, it produces exactly the same pattern of behavior about the colonies of social spiders we observed in the world.[15] We also know how variables c, r and K are involved in producing the pattern of behavior; that is, we know, for example, that "had c taken the value c_i and r taken r_n, the spider population would have instantiated the pattern of nutritional strategy $K = k_j$". Given these, a fine-grained hypothetical relationship between the model and its target system can be built:

(**HR***):

i. If M is a good model;
ii. And in M, if c takes the value c_i and r takes r_n, the spider population instantiates the pattern of nutritional strategy $K = k_j$;
iii. And in T, if c takes the value c_i and r takes r_n;
iv. Then, *hypothetically*, the spider population in T would also instantiate the pattern of nutritional strategy $K = k_j$.

Therefore, as in the first kind of inferential relationship, HRs regarding the model–world relationship can also be systematically constructed, supposing that the model is confirmed to be a good model. Given the HRs that have been constructed, we are able to answer a whole range of questions concerning the model and its target system. For example, we can not only answer the question that (i) if M is a good model; (ii) in M, if c takes the value 0.2 and r takes r_1, the spider population would instantiate the pattern of nutritional strategy $K = 0.271$; and (iii) in T, if c takes the value 0.2 and r takes r_1, then, hypothetically, (iv) the spider population in T would also instantiate the pattern of nutritional strategy $K = 0.271$, but also the question that (i) if M is a good model; (ii) in M, if c takes the value 0.7 and r takes r_1, the spider population would instantiate the pattern of nutritional strategy $K = 0.91$; and (iii) in T, if c takes the value 0.7 and r takes r_1, then, hypothetically, (iv) the spider population in T would also instantiate the pattern of nutritional strategy $K = 0.91$, and so on. Moreover, the HR between the model and its target is invariant in the sense that, given certain range(s) of interventions either on the initial and/or background conditions (e.g., from spiders to burying beetles), or on the variables of the model, the same pattern of hypothetical reasoning still holds.

To summarize, a model can be explanatory because it helps the modelers to answer two kinds of questions based on two kinds of relationships: (1) dependence relationships among variables within a model and (2) HRs that constitute basis for inference from the model to its target system; that is, the modelers infer that the counterfactual dependence relationships detected in the model might be true in the target system too, on the basis of the modelers' belief that there is a kind of holistic fit between the model and its target. Note that the first kind of questions is rooted in the model while the second is based on the modelers' hypotheses rather than on the model itself.[16] Furthermore, the underlying reason why a model can be explanatory as such is that (i) a model is a structure, that is, a dependence relationship, such that it can be used to answer counterfactual dependence questions, and that (ii) the holistic fit between a model and its target system warrants the hypothetical inference from the model to its target.

Before going to the next section, it is worth mentioning one thing regarding the relationship between Bokulich's view and my view. Recall Bokulich's (2011) "justificatory step" that serves the purpose of "specifying what the domain of applicability of the model is, and showing that the phenomenon in the real world to be explained falls within that domain" (39). My discussion of how the model might be linked to its target through the hypothetical relationships can be viewed as an extension of Bokulich's justificatory step. This is because, on the one hand, the hypothetical relationships built by modelers concern aspects of the model (and its target) that fall within the intended domain of applicability of the model. This is the very reason why the modelers bother hypothesizing and entertaining these relationships. On the other hand, the hypothetical relationships not only concern the intended domain of applicability of the model in general but, more important, also manifest how particular aspects of the model within that domain might be related with its counterparts in the target. More specifically, they manifest how the set of counterfactual dependence relationships, that is, the particular aspects of the model within its domain of applicability, can be extrapolated to its counterparts in the target. This second dimension of the justificatory step, which renders my view distinct from Bokulich's, is what has not been fleshed out in her view (recall that another distinctive feature of my view is the claim that it is the holistic fit between the model and its target that facilitates the hypothetical inference that the set of dependence relationships in the model may be extrapolated to its target).

7.5 What Is Additional in the Holistic View?

One may argue that my holistic view says nothing more than what Bokulich's (2011) view says, since both our views can come down to the simple claim that a model can be explanatory because "the elements of the model correctly capture the pattern of counterfactual dependence of the target system" (39). This is true, and for this reason, my holistic view can also be grouped into the category of structural account of explanation.

Yet, on closer inspection, one can find the merits of the holistic approach that Bokulich's view lacks (or at least it remains uncertain how it could have). In particular, the holistic view is general in scope in that it can accommodate various kinds of models and thus fares better in shedding light on modeling practice. Remember that the holistic view holds that the model bears a holistic fit to its target, which can take many different forms depending on different modeling practices. It can be strict isomorphism at one extreme, simple fitting of outputs (or patterns) at the other extreme or something in between. This is due to the fact that the holistic view treats model structure as a continuum, ranging from the strict structure that can be mapped to its target in a formal fashion (e.g., isomorphic, partial isomorphic, or the like), through the idealized or distorted structure that merely preserves certain parts or aspects of its target, to the nominal structure that might have nothing in common with the real structure of its target. On the other hand, it remains unclear how Bokulich's view could accommodate these various kinds of models and modeling, though one thing might be quite clear that her view differs from the semantic view.

As an indication of the scope of the holistic approach, consider how it could accommodate a predictive model that only concerns prediction accuracy. The model is taken from Kuhn and Johnson (2013, 19–24). Suppose, for the purpose of environmental protection, the Australian government endeavors to build a model to predict fuel efficiency. Among many vehicle characteristics, the government selects engine displacement (the volume inside the engine cylinders) as the key predictor variable, and highway miles per gallon (mpg) of the car as the key response variable. The government mainly collects data for 2010–2011 model-year cars, with 2010 having 1,107 vehicles tested and 2011 having 245 vehicles tested. The vehicles include a wide range of cars, such as buses, vans, trucks, SUV cars, sedans, sports cars, and the like. Figure 7.2 shows the plots given the data.

It can be seen from the figure that "as engine displacement increases, the fuel efficiency drops regardless of year. The relationship is somewhat linear but does exhibit some curvature towards the extreme ends of the displacement axis" (ibid., 19). Based on this roughly linear relationship, the government is able to predict that, no matter what kinds of cars are tested and no matter which year is considered, the fuel efficiency decreases as engine displacement increases.[17] Furthermore, looking closely to the data obtained, a quantitative relationship can be built between the two variables, e.g., *efficiency* = 63.2 – 11.9 × *displacement* + 0.94 × *displacement*^{2} (ibid., 22). Now the government is able to predict not only the rough trend but also the trend with accuracy.

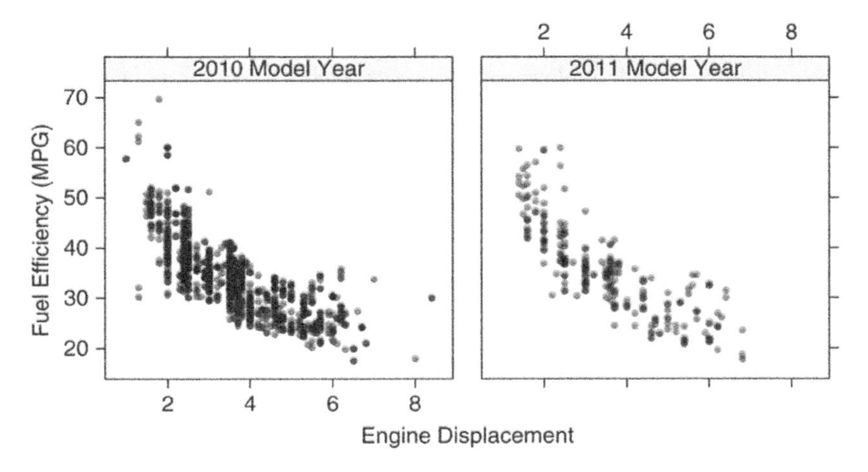

Figure 7.2 The relationship between the two variables in the model: engine displacement and fuel efficiency.

Sources: This figure comes from Kuhn and Johnson (2013, 20). Figure used with permission.

The message we get from this example is that, since the predictive model is insensitive to which particular kind of car is involved, the model is not dependent on any particular underlying structure (i.e., the working mechanism of cars) in order to give rise to the precise prediction. In other words, even if the model does not bear any formal relationship (such as partial isomorphism) to its target(s), it is still a very good model in terms of prediction precision.[18]

As another indication of the scope of the holistic approach, consider how it could accommodate a different kind of model built for the purpose of explanation rather than for prediction, namely, a minimal model drawn from Batterman and Rice (2014)—Fisher's sex ratio model:

> Fisher's model relies on a key trade-off between the ability to produce sons and daughters. This trade-off—which economists refer to as the substitution cost—tells us how many sons can be produced if one less daughter is produced. In Fisher's explanation of the 1:1 sex ratio, this substitution cost is perfectly linear—males and females cost the same amount of resources to produce, and so one fewer son means one more daughter and vice versa. This is why his model predicts a 1:1 sex ratio. Indeed, the key to deriving the 1:1 sex ratio is that natural selection will lead to equal investment in males and females, and males and females cost the same amount of resources to rear to reproductive age.
>
> (367)

It seems that the explanatory power of Fisher's model stems from the fact it shows that many different species (or populations) exhibit a common feature: the substitution cost. That is, due to the uncovering of this common feature, the model is able to explain why different species show the same 1:1 sex ratio. If this is the whole story about why Fisher's model can be explanatory, then it seems the isomorphism or partial isomorphism approach gains the victory—for sharing explanatorily relevant features is part of any mapping account of models. However, Batterman and Rice argue that the explanatory power of Fisher's model comes from somewhere else:

> Citing the features the model has in common with real systems fails to provide a satisfactory answer. Moreover, simply noting that those common features are relevant to the explanandum fails to tell us why those features are common and relevant. The answer to these questions, we contend, is that Fisher's model is a minimal model within the same universality class as the actual systems whose behavior it purports to explain. The key to understanding the explanatory power of the model is to look to the means by which the universality class is delimited and to the demonstration that Fisher's model is in that class.
>
> (ibid., 369)

The key to their argument is that, even if we explain the 1:1 sex ratio by citing the fact that the model shares the same substitution cost with its target systems, there remains the question of why different species with substantially different underlying details (e.g., mating strategies) all instantiate the same substitution cost (ibid., 374). To put it differently, the substitution cost is not the answer to our question but is the very question we try to ask, Why the substitution cost is instantiated by so different species? To answer this further question, we need a different story. In particular, we need to delineate the universality class within which the model finds itself, and this "involves the demonstration that the details that genuinely distinguish the different populations (sexual vs. asexual reproduction, etc.) are largely irrelevant for their displaying the observed 1:1 sex ratio pattern" (ibid., 374).[19] In doing so, we not only explain why there is a common 1:1 sex ratio in different species but also explain why the substitution cost that matters to the 1:1 sex ratio is so commonly instantiated by these different species

Therefore, the explanatory story regarding Fisher's model is not simply that the model shares a common feature with its target systems, but rather that the model "delimits the universality class and guarantees a kind of robustness or stability of the continuum behaviors of interest under rather dramatic changes in the lower-scale makeup of the various systems and models" (ibid., 364). In other words, the model explains why a species behaves in a certain way by showing that the species in question falls into a certain universality class. Note that Batterman and Rice argue explicitly against any *morphism* approach to models (e.g., isomorphism or partial isomorphism) because, according to them, the explanatory power of the minimal model does not come from its sharing certain features with its target system (ibid., 350).[20] I agree with Batterman and Rice, because my holistic approach also holds that in the case of minimal models (and in the case of predictive models), there is no formal relationship such as isomorphism or partial isomorphism between the model and its target system. Instead, all we have is the sharing of outputs or overall patterns, a scenario different from what the isomorphism or partial isomorphism approach requires: sharing an underlying structure. Yet, my approach also holds that there might be other kinds of models that bear such a formal relationship to their target systems. Recall that in Chapter 6 I claimed that in certain limited cases some models may bear such a formal relationship to their targets, for example, some purely mathematical models (see Baron et al. (2017).

Hence, insofar as the minimal model is explanatory, it is not a case in which the model bears an isomorphic (or partially isomorphic or the like) relationship to its target systems. Therefore, we have seen both a predictive model and an explanatory model that do not have the required strict relationships (e.g., partial isomorphism) with their targets. That being said, the situation confronting us shows a virtue of, rather than creating a problem for, the holistic approach. Recall that the holistic view treats model structure as a continuum, ranging from the strict structure through the idealized or distorted structure to the nominal structure. A nominal structure is one that might have nothing in common with the real structure of its target, and this is the scenario we have seen in the foregoing predictive and explanatory modeling. Yet, it does not follow that a nominal structure can be anything whatsoever. As discussed in the previous chapter, the minimal requirement for a structure is that it is at least a dependence relationship (causal or noncausal) that can be employed to answer Woodward's w-questions. In the case of predictive models, the dependence relationship comes down to the input-and-output relationship, and a good predictive model is one that, when an input is entered into the model, the model produces (or predicts) an output with the required accuracy (recall the dependence relationship between engine displacement and fuel efficiency in the predictive model discussed earlier). In the case of minimal models, the dependence relationship is built within the pattern of behavior, for example, in Fisher's sex ratio model, the dependence relationship may take the form that if a species were to have more sons than daughters then the species would have the tendency to produce more daughters than sons until the sex ratio attains 1:1.

In these cases the model–world relationship is still a holistic matter because we consider the model (and its target) as a whole. For example, in the predictive model case, we consider prediction precision of the model regardless of the underlying details of the model, and the prediction is not produced *additively* by several independent elements of the model but produced *interactively* by elements that connect with one another directly or indirectly. The precision of the prediction is measured by comparing the output of the model with the output of its target as a whole. In the minimal model case, the overall pattern manifested by different systems is the only concern, which is also produced interactively by elements that connect with one another directly or indirectly, although the ways of producing the same pattern might be substantially different in different systems due to disparate underlying microstructures. In testing the goodness-of-fit of the minimal model, the focus is on the degree to which the model produces the same pattern of behavior we observed in the target system, and this can be achieved in a holistic manner using certain testing methods as shown in the previous chapter.

To sum up, the merit of the holistic view is that it is general in scope such that it is able to accommodate different kinds of models and modeling practice, such as predictive and minimal modeling.

7.6 Conclusion

Woodward's view that a generalization can be explanatory because it can answer what-if-things-had-been-different questions has made a great contribution to the philosophy of scientific explanation. In the context of scientific modeling, Woodward's idea has been applied and further developed by Alisa Bokulich. Bokulich's structural account of model explanation rightly captures the intuition that a model can be explanatory because it represents the pattern of counterfactual structure of its target system, although it does not fully explain how the model can do so. I claimed that to properly account for why a model can do so, we need an adequate account of the model–world relationship. Here my holistic view of the model–world relationship entered. The underlying idea is that it is the holistic fit between the model and its target that facilitates the hypothetical inference from the model to its target, which enables us to answer Woodward's w-questions about the target in terms of the counterfactual structure of the model.

More specifically, I claimed that for a model to be explanatory it must help the modelers to answer two kinds of questions: dependence questions (corresponding to Woodward's w-questions) concerning the model itself, and hypothetical questions concerning the relationship between the model and its target system. With this construal, I further suggested that the reason why a biological model can answer these two kinds of questions is due to the fact that (1) a model is a structure, that is, a set of dependence relationships that can be employed to answer the first kind of question, and that (2) the holistic fit between the model and its target warrants the hypothetical inference from the model to its target and thus helps answer the second kind of question.

Notes

1 Other accounts of model explanation that can be found in the literature are McMullin (1978, 1984, 1985), Batterman (2002b), Elgin and Sober (2002), Craver (2008), Strevens (2008), and others.
2 Since the debate is not over whether laws of nature can be explanatory but over whether (and how) non-laws can be explanatory, I simply take it for granted that laws are explanatory.
3 Woodward (2013a) admits that the question of whether his interventionist theory can be applied to fundamental physics remains unresolved at present.
4 I sometimes use *relationship* and sometimes use *generalization* throughout the chapter. For me, a relationship is something in the world while a generalization is a linguistic entity (or a model) used to describe or summarize something in the world, for example, a relationship. I thank Pierrick Bourrat and Arnaud Pocheville for helping me make this distinction explicit.
5 In total, Woodward (1997) distinguishes three senses of invariance: (1) functional form invariance, meaning that "if a single equation or system of equations describes an invariant set of relationships then the equations themselves—their functional form and the coefficients occurring in them—should be invariant under (some range of) changes in the values of the variables occurring on the right hand side of each equation" (S36); (2) coefficient invariance, meaning that "it should be possible to intervene to change each of the coefficients in these equations separately without changing any of the other coefficients" (ibid., S36); and (3) invariance under different background conditions: the

relationship keeps invariant under varying background conditions (ibid., S33). I thank Pierrick Bourrat for bringing these differences to my attention. Pocheville et al. (2017) suggest that the third sense of Woodward's invariance involving background conditions should be labeled as "stability" and that, although "stability" is a desirable property, it is not part of the definition of invariance. Thus, they suggest that invariance and stability be clearly distinguished.

6 Note that "any generalization, no matter how accidental, will be stable under some changes in background conditions" (Woodward 1997, S32). For example, I may get up at 7 each morning regardless of what the weather is. So what Woodward means here is stronger than this, for he requires invariance under changes that are interventions. See Woodward (1997, S32) for details.

7 This example comes from Woodward (2000, 208).

8 The reader may find that what I demonstrate in this chapter about biological models can be generalized to scientific models in general. Although I think this is true, I restrict myself to the claim in the following that it at least shows how biological models can be explanatory.

9 "Agent-based modeling (ABM) is a computational modeling paradigm that enables us to describe how any agent will behave" (Wilensky and Rand 2015, 22). And by the word *agent*, "we mean an autonomous individual element of a computer simulation. These individual elements have properties, states, and behaviors" (ibid., 22).

10 This is an open-access article and permits unrestricted use of its content provided that the original authors and source are credited.

11 Note that this view is consistent with my ontic-oriented epistemic conception of scientific explanation discussed in Chapter 3, for the view insists that the explanatory power of the model is essentially related to how the model is connected to its target system in the world.

12 Other authors have the similar inferential account of the model-world relationship, for example, Suárez (2004, 2015, 2016), Bueno and Colyvan (2011), Bueno and French (2012), and Baron et al. (2017). For details of these views, see Note 32 of Chapter 6.

13 I thank Arnaud Pocheville for alerting me to this.

14 I thank Arnaud Pocheville and Pierrick Bourrat for alerting me to notice that the relationship between the model and its target system is only hypothetical rather than something associated with counterfactuals. In early versions of this chapter I treated the relationship as counterfactual, that is, I claimed that "if *M* is a good model, then had *M* behaved in such and such a way, then its target system would have also behaved in such and such a way". I also thank Arnaud Pocheville for alerting me to know that the inference from the model to its target (and vice versa) is, in fact, a hypothesis: because the model behaves in such and such a way we hypothesize that the target would also behave in such and such a way. The formation of the HR statement is indebted to many colleagues, including Arnaud Pocheville, Paul Griffiths, Pierrick Bourrat, and Brian Hedden.

15 The goodness-of-fit of a model is always dependent on the goals of modeling that modelers have in mind. Here I assume that the goal of the agent-based modeling is to generate the *same* pattern of behavior about the colonies of social spiders we observed in the real world; the degree of *sameness* between the model and its target system can be measured using certain mathematical tools (e.g., the maximum likelihood estimation method).

16 I thank Wendy Parker for helping me to make it clear that the second kind of questions is not based on the model itself but on the modelers' hypotheses which aim to extrapolate results of the model to its target system.

17 The actual relationship between engine displacement and fuel efficiency is more complicated than what has been presented here. But for illustration purposes, I simply focus on the key features of this example and leave the details to the reader who is interested in the actual relationship.

18 One might be worried that this is an irrelevant case for arguing against the semantic view, because the model is about the relationship between engine displacement and efficiency, rather than about the underlying mechanisms of different kinds of cars. Hence, a lack of isomorphism of the mechanism between the model and its target system is orthogonal to whether the model has an appropriate relationship between engine displacement and efficiency as its target system does. They are different questions. Even if there is isomorphism, that must be isomorphism of the displacement–efficiency relationship between the model and its target (i.e., they both have this displacement–efficiency relationship). However, if the foregoing reasoning is true, then it seems unclear how an isomorphic relationship between the model and its target can ever arise. On one hand, it is unclear how isomorphism can be so loosely understood as a relation between two input–output structures (e.g., displacement–efficiency). On the other, even if isomorphism can be so loosely understood, there remains a question of why we should keep using the concept "isomorphism" given that the model being discussed can be readily treated within my holistic framework (deflationarily construed). I thank Patrick McGivern for raising this concern for me.

19 *Universality* refers to the fact that "distinct systems can in certain cases display the same type of behavior" (Batterman 2002a, 23), and the *universality class* refers to the class of systems that all display the same type of behavior. For details of universality and the universality class see Batterman (2002a, 2002b, 2010).

20 One may wonder if it is possible that some extremely abstract minimal model still bears some kind of morphism relationship to its physical systems. However, as Batterman argues extensively, many minimal models, due to the use of the method of the renormalization group and the use of limiting asymptotic idealizations, are not even *representative* with respect to their target systems (i.e., they are not representations of their target systems); hence, it is hard to find any morphism relationship between the model and its target system (Batterman 2010, 18–19; Batterman and Rice 2014, 350; also see Batterman 2002a, 2002b, 2009).

8 Conclusion

In the 1960s, J. J. C. Smart (1963) famously said that

> [f]rom a logical point of view biology is related to physics and chemistry in the way in which radio-engineering is related to the theory of electromagnetism, etc.... Just as the radio-engineer uses physics to explain why a circuit with a certain wiring diagram behaves as it does, so the biologist uses physics and chemistry to explain why organisms or parts of organisms (e.g., cell nuclei), with a certain natural-history description, behave as they do.
>
> (57)

For Smart, biology is (or at least resembles) a sort of engineering science that does not have its own autonomy, since we must appeal to somewhere else such as physics or chemistry so as to explain biological phenomena. This view was popular among Smart's contemporaries, for example, Francis Crick (1966), a founder of modern genetics, once claimed that "the ultimate aim [of science] ... is in fact to explain all biology in terms of physics and chemistry" (10; cf. Mayr 1996, 98).

This book has shown that the biological sciences do have their own autonomy, especially their explanatory autonomy. In this concluding chapter, let me recapitulate the main aspects regarding why I think the biological sciences have their explanatory autonomy. The problem of explanatory autonomy concerns how the biological sciences can be explanatory, which in turn relates to topics such as reductionism, laws of nature, and biological models. So the book starts with considering reductionism. Reductionism has many forms, and those related to the autonomy problem are metaphysical and explanatory reductionism. One classical way of raising issues with metaphysical reductionism is through the multiple realization (MR) thesis. Based on Polger and Shapiro's criterion for MR, I have developed a multiple mechanistic realization (MMR) thesis. In developing the MMR thesis, I have two purposes in mind: (a) showing that the phenomenon of MMR is widespread in the biological world and (b) showing that, even if the MMR thesis holds, it, on its

DOI: 10.4324/9781003148029-8

own, neither demonstrates nor refutes metaphysical and (more important) explanatory reductionism. I suggest that, to take sides in the dispute surrounding explanatory reductionism, one needs an account of scientific explanation.

So, Chapter 3 attempts to develop an (ontic-oriented) epistemic conception of scientific explanation that incorporates epistemic considerations relating to contextual elements, extra information, and pragmatic issues. There are two ways to take into account contextual elements: (a) the context-dependence side: depending on different contextual factors (cellular, organelles, tissues, organismic or even environmental), a molecular entity can give rise to many, sometimes even radically distinct, effects and (b) the context-independence side: a higher-level phenomenon can be partially independent of its lower-level underpinnings, because changing the lower-level underpinnings from one to another does not necessarily result in changes to the higher-level phenomenon. The second side reminds us of the MMR thesis. The claim is that the MMR thesis can be employed in the context of scientific explanation as a foundation for arguing against reductionism, because the thesis guarantees the partial independence of the higher-level phenomenon. In the context of scientific explanation, this entails that explanation can proceed at a level of analysis without making an appeal to its underlying happenings. Lumping them together, the two sides show that explanatory reduction does not always happen in practice.

The extra-information dimension concerns the fact that explanations in many areas of biology (e.g., developmental biology, ecology, systems biology, etc.) cannot be achieved without the supplement of knowledge from domains other than molecular biology. This kind of knowledge may include, but is not restricted to, spatial, structural, geometrical or topological information that cannot be simply subsumed into, or reduced to, the domain of molecular biology. The pragmatic dimension of scientific explanation points to the fact that there is no better explanation *simpliciter*, because explanations are typically advanced for the purpose of answering specific questions being asked in certain contexts. Hence, it is not always the case that we take a reductive strategy and privilege micro-level over macro-level explanations. On the contrary, we sometimes have good reasons to privilege macro-level over micro-level explanations. Putting together, these considerations demonstrate that explanations in the biological sciences do not always operate in a reductionist fashion.

The second topic connected with the explanatory autonomy of the biological sciences concerns laws of nature, which traditionally are believed to be the only legitimate vehicle of scientific explanation. I argued in Chapter 4 that we do not have lawlike items in the biological sciences we typically find in physics. The absence of laws in the biological sciences often leads philosophers to worry about how the biological sciences can be explanatory. To remedy the absence, some philosophers strive to redefine laws so as to make it possible that the biological sciences do have laws. For example, Sober and others propose that certain mathematical models count as genuine laws. However, I have shown that these proposed laws are not genuine laws. If they are laws, then many accidental facts also count as laws. An alternative approach to looking for an explanatory vehicle, I suggest, is to look at scientific practice more closely. In doing so, we find that scientists in the biological sciences usually use a wide variety of models to explain biological phenomena. These models may merely consider a single species, or a tiny mechanism found in a couple of organisms on earth; they are so localized and so restricted spatiotemporally that it is difficult to regard them as laws. Yet, to the extent that they are able to answer our how-, why- and what-questions, I think they are explanatory.

This leads us to biological models. If it is models in the biological sciences that explain, then the burden of proof now lies in how models can be explanatory. To answer how models can be explanatory, we first need to understand how a model is related to its target phenomena. That is, we need to understand the model–world relationship. One influential and popular view in the literature is the similarity view. In particular, Michael Weisberg's similarity account deserves special attention, because it is the most detailed, developed and sophisticated version of the similarity view we have seen so far. The account starts with the intuitive idea that a good model is one that resembles its target system in certain aspects and to certain degrees. To capture this idea and to quantify the nitty-gritty, Weisberg proposes a mathematical formula to measure the degree to which a model is similar to its target system. Yet, I argue that, although it looks plausible, it, in many respects, falls short of capturing key features of modeling practice. To begin with, his account is simply too abstract to capture what is going on in practice. Second, it implies an atomistic conception of model features while modeling practice often proceeds in a holistic manner. Third, Weisberg's atomistic conception of model features can be traced back to a problematic set-theoretic approach to the structure of models.

As an alternative to Weisberg's similarity view, I propose a holistic view of models and modeling. The holistic view holds that the model–world relationship constitutes a holistic fit, where *holistic fit* refers to the degree to which the model has the same structure (or output, pattern) as its target system or refers to the distance between two structures (or outputs, patterns). This holistic view does not arise from armchair contemplating but from looking closely at biological modeling practice. In particular, it comes from examining carefully two most common testing methods in modeling practice: the maximum likelihood estimation (MLE) method and the least squares estimation (LSE) method. The general lesson by examining these methods is that in practice modelers usually compare a model with its target system holistically. Modelers typically treat a model as a whole and know that it is the interacting parts of the model that produce outputs rather than each individual part that does so.

Coupled with a holistic view of models and modeling, I suggest a deflationary understanding of model structure. According to this view, model structures, of various kinds (e.g., concrete physical structures, equations, graphs, pictures, abstract structures in logic and meta-mathematics, etc.), are important *inferential tools* in modeling practice. However, although model structures can be of various kinds, they have one thing in common: they are, or at least can be described as, *dependence relationships*. A genuine dependence relationship (causal or noncausal) is one that allows one to answer Woodward's *what-if-things-had-been-different* questions (w-questions); that is, it tells us how certain variables would change if other variables were to be changed or manipulated. The deflationary view has merits that its rivals (e.g., the semantic view) lack. For example, it is able to accommodate different kinds of model structures and therefore consonant with the heterogeneous nature of modeling practice. Moreover, it is able to smoothly treat models involving idealizations, because, within the deflationary framework, a model structure can be an idealized or even distorted structure as long as it enables modelers to answer various w-questions.

Finally, on the basis of the holistic view of models and modeling, I propose a holistic view of how biological models can be explanatory. This view draws heavily on Woodward's interventionist account of scientific explanation and Bokulich's structural account of model explanation. The view is that, for a model to be explanatory, it must help the modelers to answer two kinds of questions: counterfactual dependence questions (corresponding to Woodward's w-questions) that concern the model itself, and *hypothetical* questions that concern the relationship between the model and its target system. Furthermore, I suggest that the reason a biological model can answer these two kinds of questions is that (a) a model is a structure, that is, a set of dependence relationships that can be employed to answer the first kind of question, and (b) the holistic fit between the model and its target system warrants the hypothetical inference from the model to its target and thus helps us answer the second kind of question.

To conclude, this book has shown that, without reducing to the lower-level and with no appeal to the laws of nature, the biological sciences can be explanatory because they typically use models to explain biological phenomena. Yet what this book has not shown is how noncausal models in the biological sciences can be explanatory. Recall that, although this book briefly discusses one minimal model (i.e., Fisher's sex ratio model), the focus is on causal models (e.g., the leaf gas-exchange model, the model involving the relationship between the plant's height and the amount of fertilizer the plant receives, etc.). Nowadays, there is a substantial debate in the general philosophy of science about how a noncausal model (e.g., a mathematical, optimality or minimal model) can explain physical phenomena. This problem deserves more attention by philosophers of biology. Although I believe the holistic account of model explanation presented in this book applies to noncausal models, showing how the holistic account could shed light on noncausal models must await another occasion. Moreover, model testing methods in the biological sciences are of a great many forms, but my discussion in this book is limited to two traditional and relatively simple methods (i.e., MLE and LSE). So there is a problem of how far the proposed holistic view of the model–world relationship can go. Although it is true that they are the two most commonly used testing methods in the biological sciences and many lately developed more advanced methods are either based on or related to them, they are nevertheless only part of the whole story. Hence, it would be very interesting to see if there are alternative testing methods in the biological sciences that speak against the proposed holistic view regarding the model–world relationship.

Appendix 1

The Leaf Gas-Exchange Model

Box A1.1 The structural equations for the leaf gas-exchange model

SLM $(X_1) = N(0,\sigma_1)$[1] $\qquad \varepsilon_2 = N(0,\sigma_2)$
Leaf nitrogen concentration $(X_2) = a_1X_1 + b_2\varepsilon_2 \qquad \varepsilon_3 = N(0,\sigma_3)$
Stomatal conductance $(X_3) = a_2X_2 + b_3\varepsilon_3 \qquad \varepsilon_4 = N(0,\sigma_4)$
Net photosynthetic rate $(X_4) = a_3X_3 + b_4\varepsilon_4 \qquad \varepsilon_5 = N(0,\sigma_5)$

Internal CO_2 *concentration* $(X_5) = a_4X_3 + a_5X_4 + b_5\varepsilon_5$

$$Cov(X_1,\varepsilon_2) = Cov(X_1,\varepsilon_3) = Cov(X_1,\varepsilon_4) = Cov(X_1,\varepsilon_5) = Cov(X_2,\varepsilon_3) = Cov(X_2,\varepsilon_4)$$

$$= Cov(X_2,\varepsilon_5) = Cov(X_3,\varepsilon_4) = Cov(X_3,\varepsilon_5) = Cov(X_4,\varepsilon_5) = Cov(\varepsilon_2,\varepsilon_3) = Cov(\varepsilon_2,\varepsilon_4)$$

$$= Cov(\varepsilon_2,\varepsilon_5) = Cov(\varepsilon_3,\varepsilon_4) = Cov(\varepsilon_3,\varepsilon_5) = Cov(\varepsilon_4,\varepsilon_5) = 0$$

Since some variables are causally independent of one another (e.g., X_1 and ε_4, X_2 and ε_5, and so on), the covariance between them is simply zero.

Box A1.2 Predicted population variances and covariances for variables described in Box A1.2

	X_1	X_2	X_3	X_4	X_5
X_1	$Var(X_1)$	$a_1 Var(X_1)$	$a_1 a_2 Var(X_1)$	$a_1 a_2 a_3 Var(X_1)$	$(a_1 a_2 a_4 + a_1 a_2 a_3 a_5) Var(X_1)$
X_2		$Var(X_2)$	$a_1^2 a_2 Var(X_1)$ $+a_2 b_2^2 Var(\varepsilon_2)$	$a_1^2 a_2 a_3 Var(X_1)$ $+a_2 a_3 b_2^2 Var(\varepsilon_2)$	$\left(a_1^2 a_2 a_4 + a_1^2 a_2 a_3 a_5\right) Var(X_1)$ $+\left(a_2 a_4 b_2^2 + a_2 a_3 a_5 b_2^2\right) Var(\varepsilon_2)$
X_3			$Var(X_3)$	$a_1^2 a_2^2 a_3 Var(X_1)$ $+a_3 b_3^2 Var(\varepsilon_3)$	$\left(a_1^2 a_2^2 a_4 + a_1^2 a_2^2 a_3 a_5\right) Var(X_1)$ $+\left(a_4 b_3^2 + a_3 a_5 b_3^2\right) Var(\varepsilon_3)$
X_4				$Var(X_4)$	$a_1^2 a_2^2 a_3^2 a_5 Var(X_1) + a_3 a_4 Var(X_3)$ $+a_3^2 a_5 b_3^2 Var(\varepsilon_3) + a_5 b_4^2 Var(\varepsilon_4)$
X_5					$Var(X_5)$

Note

1 $N(0,\sigma)$ means "a normally distributed random variable with a population mean of zero and a population standard deviation of σ". Because our interest is in the causal relations between variables but not the mean values of the variables themselves, all variables are "centered" by subtracting the mean value of each variable. This treatment ensures that the mean of each transformed variable is zero, and thus, the intercepts are zero. Besides, since we assume that all error variables (i.e., ε) are unit normal variables, they are with a zero mean and a standard deviation of 1 (Shipley 2002, 105–106).

Appendix 2

Derivation of Weisberg's Similarity Account From the Jaccard Similarity Coefficient[1]

One can show that Weisberg's similarity index is the weighted average of the similarity coefficients on mechanisms and attributes. The demonstration goes as follows:

$$J(A,B) = \frac{(A \cap B)}{(A \cup B)}$$

$$W(M_m,T_m) = \frac{(M_m \cap T_m)}{(M_m \cup T_m)} = J(M_m,T_m)$$

$$W(M_a,T_a) = \frac{(M_a \cap T_a)}{(M_a \cup T_a)} = J(M_a,T_a)$$

$$W(M,T) = \frac{(M_m \cup T_m)J(M_m,T_m) + (M_a \cup T_a)J(M_a,T_a)}{(M_m \cup T_m) + (M_a \cup T_a)} =$$

$$\frac{(M_m \cup T_m)\frac{(M_m \cap T_m)}{(M_m \cup T_m)} + (M_a \cup T_a)\frac{(M_a \cap T_a)}{(M_a \cup T_a)}}{(M_m \cup T_m) + (M_a \cup T_a)} = \frac{(M_m \cap T_m) + (M_a \cap T_a)}{(M_m \cup T_m) + (M_a \cup T_a)}$$

The last equation is just Weisberg's formula. If the sets of mechanisms and attributes are disjoint (which, in fact, is an assumption made by Weisberg), that is, if $M_m \cap M_a = T_m \cap T_a = M_m \cap T_a = T_m \cap M_a = \varnothing$, then Weisberg's similarity index is the Jaccard similarity index on the union of mechanisms and attributes:

$$W(M,T)=\frac{(M_m\cap T_m)+(M_a\cap T_a)}{(M_m\cup T_m)+(M_a\cup T_a)}=$$
$$\frac{((M_m\cup M_a)\cap(T_m\cup T_a))}{((M_m\cup M_a)\cup(T_m\cup T_a))}=J(M_m\cup M_a,T_m\cup T_a)$$

Proof of this equation follows directly from:

$$|A\cup B|=|A|+|B|-|A\cap B|$$

Let us prove the numerator:

$$((M_m\cup M_a)\cap(T_m\cup T_a))=$$
$$((M_m\cap T_m)\cup(M_a\cap T_m)\cup(M_m\cap T_a)\cup(M_a\cap T_a))=$$
$$(M_m\cap T_m)\cup(M_a\cap T_a)=(M_m\cap T_m)+(M_a\cap T_a)-$$
$$((M_m\cap T_m)\cap(M_a\cap T_a))=$$
$$(M_m\cap T_m)+(M_a\cap T_a)$$

The denominator:

$$(M_m\cup M_a)\cup(T_m\cup T_a)=(M_m\cup T_m)+(M_a\cup T_a)+$$
$$((M_m\cup T_m)\cap(M_a\cup T_a))=(M_m\cup T_m)+(M_a\cup T_a)$$

This means that if attributes and mechanisms are disjoint, similarity on mechanisms and attributes can be thought of somehow as independent. If mechanisms and attributes were not disjoint, however, elements in the intersection would be counted twice in the average, and the Weisberg's similarity index would be simply wrong.

Note

1 This demonstration comes from Arnaud Pocheville.

Appendix 3

Testing a Scientific Model Using the Tetrad Program[1]

Tetrad is a computer program developed by Clark Glymour, Richard Scheines and Peter Spirtes (and many others) at Carnegie Mellon University that is used to create, test and search for causal and statistical models.

Now we use this program to estimate parameters of a causal model given a hypothetical data set.[2] This model describes the causal effects of various factors on people's behavior of donation (Cryder et al. 2013). This model has five variables: tangibility condition (how detailed the situation is described to the donors), imaginability (how concrete the situation is), sympathy (how much sympathy for the target), impact (how much impact will the donors' money have on the target) and amount donated (actual amount of money donated by the donors). The hypothesized model is described in Figure A3.1.

With the data at hand, we now test this model using the Tetrad program. The testing process involves finding the most likely values for the parameters in the model, and this, in turn, involves using the maximum likelihood estimation method, as described in Chapter 5. It can be expected that the model will fit the data very well, because the data are just generated by this model. The result of the testing is shown in Figure A3.2.

The model's degrees of freedom are 4, and the chi-square test of this model assumes that the maximum likelihood fitting function over the measured variables has been minimized—that is, it assumes that we have found the most likely values for the parameters that can minimize the difference between the predicted covariance matrix and observed covariance matrix. The value of the chi-square test is 3.0327, and the probability of having observed the measured minimum difference (assuming that the predicted and observed covariances are identical except for random sampling variation) is 0.5524, which is significantly larger than the threshold value 0.05. Therefore, the hypothesized model is not to be rejected given the data.

Remember that our purpose for discussing this model using the Tetrad program is to show how the free parameters in a model might interact with one another in producing a best fit of the model to its target system.

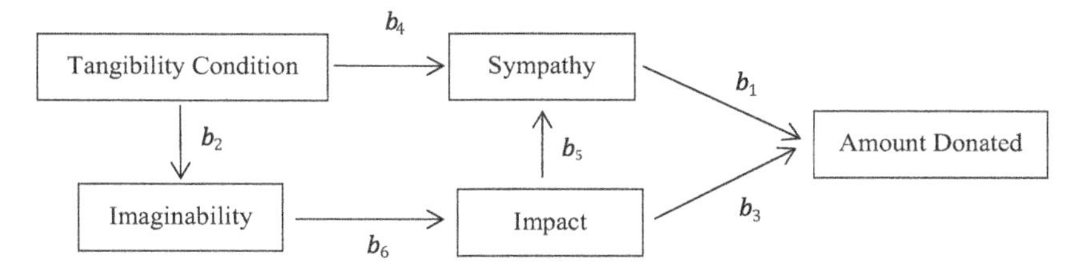

Figure A3.1 A causal model among five variables. For purpose of illustration, error variables are not shown in the figure. b_i denotes free parameters to be estimated.

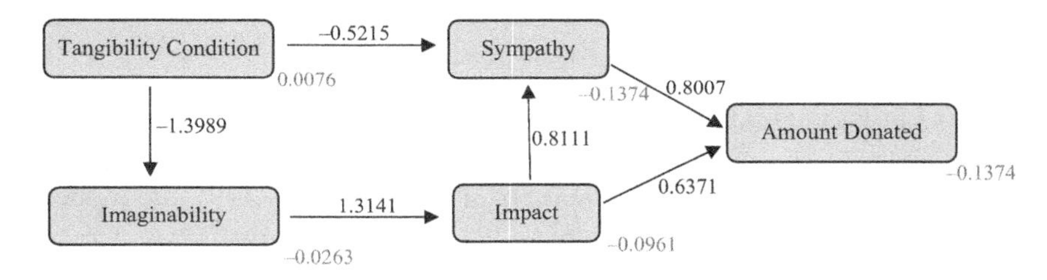

Figure A3.2 Results of using the Tetrad program to estimate the free parameters in the causal model. Numbers in black show the values of the parameters, and numbers in green show the mean values of the variables.

To show this, let me perform a little operation on the model: now we change the value of the parameter along the path linking Sympathy to Amount Donated from 0.8007 to 0.7200; that is, we fix the value of this parameter at 0.7200. The other parameters are still free parameters because they are not fixed. Then we test this modified model given the same data set we just used above. The result is shown in Figure A3.3.

It can be seen from Figure A3.3 that, by changing the value of the parameter along the path linking Sympathy to Amount Donated from 0.8007 to 0.7200, the values of other parameters have also been changed. For example, the value of the parameter linking Impact to Amount Donated has been remarkably changed from 0.6371 to 0.7163, the value of the parameter linking Tangibility Condition to Sympathy has been changed from −0.5215 to −0.5021 and so on.

Interestingly, given these changes, the modified model still fits its target system (i.e., the data set) to an extent. In this modified model, the chi-square test also assumes that the maximum likelihood fitting function over the measured variables has been minimized. With the degrees of freedom being 5 (since we fix one parameter, there is one less parameter to estimate and thus one more degree of freedom), the result of the chi-square test is 9.0028, and the probability of having observed the

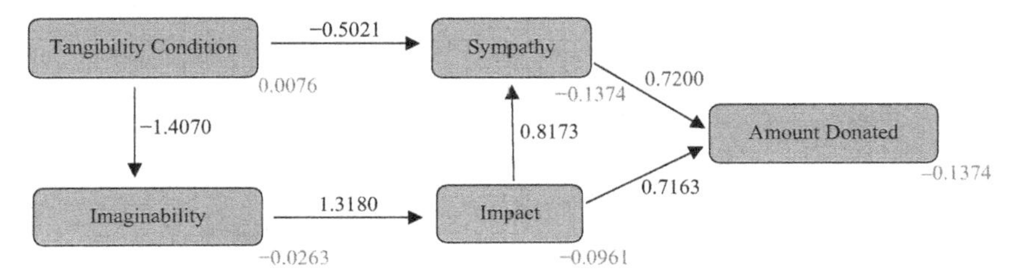

Figure A3.3 Results of using Tetrad to estimate the free parameters in the causal model by fixing the value of one parameter. The number in red is the value that has been fixed.

measured minimum difference is 0.1090, which is larger than the threshold value 0.05. Therefore, the modified model is also not to be rejected given the data.

The moral of this discussion is twofold. First, the act of choosing values for one parameter does have an impact on choosing values for the other parameters in the model. Second, although one parameter in a model might be changed significantly (e.g., from 0.8007 to 0.7200), the model may still fit its target system by adjusting the other parameters in a way that produces a best combination of values of the parameters given the data, that is, produces a best fit of the model to its target system.

Notes

1 I thank Paul Griffiths and Arnaud Pocheville for drawing my attention to the Tetrad program and its relevance to my arguments presented in Chapter 6

2 For illustration purposes, the data set is generated by the Tetrad program based on the hypothesized model. We have one sample, and the sample size is 1,000. Although the data set does not come from the real world, the following discussion shows that our conclusion does not depend on where the data set comes from.

References

Aizawa, K. (2007). "The Biochemistry of Memory Consolidation: A Model System for the Philosophy of Mind." *Synthese*, 155 (1): 65–98.

Aizawa, K. (2009). "Neuroscience and Multiple Realization: A Reply to Bechtel and Mundale." *Synthese*, 167 (3): 493–510.

Aizawa, K. (2013). "Multiple Realization by Compensatory Differences." *European Journal for Philosophy of Science*, 3 (1): 69–86.

Aizawa, K., and Gillett, C. (2009a). "Levels, Individual Variation, and Massive Multiple Realization in Neurobiology." In J. Bickle (Ed.), *Oxford Handbook of Philosophy and Neuroscience*, 539–582. Oxford: Oxford University Press.

Aizawa, K., and Gillett, C. (2009b). "The (Multiple) Realization of Psychological and other Properties in the Sciences." *Mind and Language*, 24 (2): 181–208.

Aizawa, K., and Gillett, C. (2011). "The Autonomy of Psychology in the Age of Neuroscience." In P. M. Illari, F. Russo, and J. Williamson (Eds.), *Causality in the Sciences*, 202–223. Oxford: Oxford University Press.

Akaike, H. (1974). "A New Look at the Statistical Model Identification." *IEEE Transactions on Automatic Control*, 19 (6): 716–723.

Alberch, P., and Gale, E. A.. (1983). "Size Dependence during the Development of the Amphibian Foot. Colchicine-Induced Digital Loss and Reduction." *Journal of Embryology and Experimental Morphology*, 76 (1): 177–197.

Alberch, P., and Gale, E. A.. (1985). "A Developmental Analysis of an Evolutionary Trend: Digital Reduction in Amphibians." *Evolution*, 39 (1): 8–23. doi:10.2307/2408513.

Allman, E. S., and Rhodes, J. A. (2004). *Mathematical Models in Biology: An Introduction*. Cambridge: Cambridge University Press.

Andersen, H. (2012). "The Case for Regularity in Mechanistic Causal Explanation". *Synthese*, 189 (3): 415–432.

Andersen, H. (2018). "Complements, Not Competitors: Causal and Mathematical Explanations." *British Journal for the Philosophy of Science*, 69 (2): 485–508.

Anderson, J. A. (1995). *An Introduction to Neural Networks*. Cambridge: MIT Press.

Armstrong, D. (1978). *A Theory of Universals*. Cambridge: Cambridge University Press.

Armstrong, D. (1983). *What Is a Law of Nature?* Cambridge: Cambridge University Press.

Armstrong, D. (1991). "What Makes Induction Rational?" *Dialogue*, 30: 503–511.

Armstrong, D. (1993). "The Identification Problem and the Inference Problem", *Philosophy and Phenomenological Research*, 53: 421–422.
Army Corps of Engineers. (1963). *Technical Report on Barriers: A Part of the Comprehensive Survey of San Francisco Bay and Tributaries, California. Appendix H, Volume 1: Hydraulic Model Studies*. San Francisco: Army Corps of Engineers.
Arora, K. K., Filburn, C. R., and Pedersen, P. L. (1993). "Structure/Function Relationships in Hexokinase." *The Journal of Biological Chemistry*, 268 (24): 18259–18266.
Ayala, F. J. (1968). "Biology as An Autonomous Science." *American Scientist*, 56: 207–221.
Baker, A. (2005). "Are There Genuine Mathematical Explanations of Physical Phenomena?" *Mind*, 114 (454): 223–238.
Baker, A. (2009). "Mathematical Explanation in Science". *British Journal for the Philosophy of Science*, 60 (3): 611–633.
Baker, A., and Colyvan, M. (2011). "Indexing and Mathematical Explanation". *Philosophia Mathematica*, 19 (3): 323–334.
Balari, S., and Lorenzo, G. (2014). "Ahistorical Homology and Multiple Realizability." *Philosophical Psychology*, 28 (6): 881–902.
Barabási, A.-L., and Oltvai, Z. N. (2004). "Network Biology: Understanding the Cell's Functional Organization." *Nature Reviews Genetics*, 5 (2): 101–113.
Baron, S., Colyvan, M., and Ripley, D. (2017). "How Mathematics Can Make a Difference." *Philosophers' Imprint* 17 (3): 1–29.
Batterman, R. (2000). "Multiple Realizability and Universality". *British Journal for the Philosophy of Science*, 51 (1): 115–145.
Batterman, R. W. (2002a). "Asymptotics and the Role of Minimal Models." *The British Journal for the Philosophy of Science*, 53 (1): 21–38.
Batterman, R. W. (2002b). *The Devil in the Details: Asymptotic Reasoning in Explanation, Reduction, and Emergence*. Oxford: Oxford University Press.
Batterman, R. W. (2009)." Idealization and Modelling." *Synthese*, 169 (3): 427–446.
Batterman, R. W. (2010). "On the Explanatory Role of Mathematics in Empirical Science." *The British Journal for the Philosophy of Science*, 61 (1): 1–25.
Batterman, R. W. (2014). "Reduction and Multiple Realizability". *PhiSci-Archive*, http://philsci-archive.pitt.edu/11190/.
Batterman, R. W., and Rice, C. C. (2014). "Minimal Model Explanations." *Philosophy of Science*, 81 (3): 349–376.
Beatty, J. H. (1995). "The Evolutionary Contingency Thesis". In E. Sober (Ed.), *Conceptual Issues in Evolutionary Biology*, 217–247. Cambridge: MIT Press.
Bechtel, W. (2007). "Reducing Psychology While Maintaining Its Autonomy via Mechanistic Explanations". In M. Schouten, and H. L. de Jong (Eds.), *The Matter of the Mind: Philosophical Essays on Psychology, Neuroscience and Reduction*. New York: Blackwell Publishing.
Bechtel, W. 2008. *Mental Mechanisms: Philosophical Perspectives on Cognitive Neuroscience*. New York: Taylor & Francis Group.
Bechtel, W. 2009. "Looking Down, Around, and Up: Mechanistic Explanation in Psychology." *Philosophical Psychology*, 22 (5): 543–564.
Bechtel, W., and Abrahamsen, A. (2005). "Explanation: A Mechanist Alternative". *Studies in History and Philosophy of Biological and Biomedical Science*, 36 (2): 421–441.

Bechtel, W., and Mundale, J. (1999). "Multiple Realizability Revisited: Linking Cognitive and Neural States." *Philosophy of Science*, 66 (2): 175–207.

Bekker, P. A., Merckens, A., and Wansbeek, T. J. (2014). *Identification, Equivalent Models, and Computer Algebra: Statistical Modelling and Decision Science*. Cambridge: Academic Press.

Berg, J. M., Tymoczko, J. L., and Stryer, L. (2002). "Isozymes Provide a Means of Regulation Specific to Distinct Tissues and Developmental Stages (Section 10.3)." In *Biochemistry*, 5th edition. New York: W. H. Freeman.

Berryman, A. A. (2003). "On Principles, Laws and Theory in Population Ecology." *Oikos*, 103: 695–701.

Bickle, J. (2003). *Philosophy and Neuroscience: A Ruthlessly Reductive Account*. Dordrecht: Kluwer.

Bickle, J. (2010). "Has the Last Decade of Challenges to the Multiple Realization Argument Given Aid and Comfort to Psychoneural Reductionists." *Synthese*, 177 (2): 247–260.

Bickle, J. (2013). "Multiple Realizability." In *Stanford Encyclopedia of Philosophy*. http://plato.stanford.edu/entries/multiple-realizability/.

Bird, A. (2001). "Necessarily, Salt Dissolves in Water". *Analysis*, 61 (4): 267–274.

Bird, A. (2002). "On Whether Some Laws are Necessary". *Analysis*, 62 (3): 257–270.

Bird, A. (2004). "Strong Necessitarianism: The Nomological Identity of Possible Worlds". *Ratio*, 17 (3): 256–276.

Bird, A. (2005a). "Laws and Essences". *Ratio*, 18 (4): 437–461.

Bird, A. (2005b). "The Dispositionalist Conception of Laws". *Foundations of Science*, 10 (4): 353–370.

Block, N., and Fodor, J. (1972). "What Psychological States Are Not," *Philosophical Review*, 81: 159–181.

Bokulich, A. (2008). *Reexamining the Quantum-Classical Relation*. Cambridge: Cambridge University Press Cambridge.

Bokulich, A. (2011). "How Scientific Models Can Explain." *Synthese*, 180 (1), 33–45.

Bokulich, A. (2012). "Distinguishing Explanatory from Nonexplanatory Fictions." *Philosophy of Science*, 79 (5), 725–737.

Bokulich, A. (2013). "Explanatory Models Versus Predictive Models: Reduced Complexity Modelling in Geomorphology." In V. Karakostas, and D. Dieks (Eds.), *EPSA11 Perspectives and Foundational Problems in Philosophy of Science*, 115–128. Cham: Springer International Publishing.

Bokulich, A. 2016. "Fiction as a Vehicle for Truth: Moving beyond the Ontic Conception." *The Monist*, 99 (3): 260–279.

Boveri, T. (1914). *Zur Frage der Entstehung maligner Tumoren*. Jena: G. Fischer.

Boyd, R. (1980). "Materialism without Reductionism: What Physicalism Does Not Entail." In N. Block (Ed.), *Readings in Philosophy of Psychology* (Vol. 1), 1–67. Cambridge: Harvard University Press.

Boyd, R. (1991). "Realism, Anti-Foundationalism and the Enthusiasm for Natural Kinds." *Philosophical Studies*, 61 (1/2): 127–148.

Boyd, R. (1999). "Homeostasis, Species, and Higher Taxa." In R. A. Wilson (Ed.), *Species: New Interdisciplinary Essays*, 141–185. Cambridge: MIT Press.

Brandon, R. N. (1997). "Does Biology Have Laws? The Experimental Evidence". *Philosophy of Science*, 64 (4): S444–S457.

Brandon, R. N. (2006). "The Principle of Drift: Biology's First Law." *The Journal of Philosophy*, 103 (7): 319–335.

Brigandt, I. (2010). "Beyond Reductionism and Pluralism: Toward an Epistemology of Explanatory Integration in Biology." *Erkenntnis*, 73 (3): 295–311.

Brigandt, I., and Griffiths, P. (2007). "The Importance of Homology for Biology and Philosophy." *Biology and Philosophy*, 22 (5): 633–641.

Brigandt, I., and Love, A. (2015). "Reductionism in Biology." In E. N. Zalta (Ed.), *The Stanford Encyclopedia of Philosophy* (Fall 2015 edition), http://plato.stanford.edu/archives/fall2015/entries/reduction-biology/.

Bueno, O. (1997). "Empirical Adequacy: A Partial Structures Approach". *Studies in History and Philosophy of Science (Part A)*, 28 (4): 585–610.

Bueno, O., and Colyvan, M. (2011). "An Inferential Conception of the Application of Mathematics". *Noûs*, 45 (2): 345–374.

Bueno, O., and French, S. (2012). "Can Mathematics Explain Physical Phenomena?" *British Journal for the Philosophy of Science*, 63 (2012): 85–113.

Bueno, O., French, S., and Ladyman, J. (2002) "On Representing the Relationship between the Mathematical and the Empirical." *Philosophy of Science*, 69 (3): 497–518.

Bueno, O., French, S., and Ladyman, J. (2012). "Empirical Facts and Structure Transference: Returning to the London Account". *Studies in History and Philosophy of Modern Physics*, 43 (2012): 95–104.

Burian, R. M. (2004). "Molecular Epigenesis, Molecular Pleiotropy, and Molecular Gene Definitions". *History and Philosophy of the Life Sciences*, 26 (1): 59–80.

Byerly, H. (1990). "Causes and Laws: The Asymmetry Puzzle." *PSA: Proceedings of the Biennial Meeting of the Philosophy of Science Association*, 1990: 545–555.

Callender, C., and Cohen, J. (2006). "There is No Special Problem about Scientific Representation". *Theoria*, 21 (1): 67–85.

Cartwright, N. (1983). *How the Laws of Physics Lie*. Oxford: Oxford University Press.

Cartwright, N. (1989). *Nature's Capacities and Their Measurement*. Oxford: Oxford University Press.

Cartwright, N. (1999). *The Dappled World: A Study of the Boundaries of Science*. Cambridge: Cambridge University Press.

Charnov, E. (1982). *The Theory of Sex Allocation*. Princeton: Princeton University Press.

Clapp, L. (2001). "Disjunctive Properties: Multiple Realizations." *Journal of Philosophy* 98 (3): 111–136.

Colyvan, M. (2001). *The Indispensability of Mathematics*. Oxford: Oxford University Press.

Colyvan, M. (2002). "Mathematics and Aesthetic Considerations in Science". *Mind*, 111 (441): 69–74.

Colyvan, M. (2010). "There is No Easy Road to Nominalism". *Mind*, 119 (474): 285–306.

Colyvan, M. (2012). *An Introduction to the Philosophy of Mathematics*. Cambridge: Cambridge University Press.

Colyvan, M., and Ginzburg, L. R. (2003). "Laws of Nature and Laws of Ecology." *Oikos*, 101: 649–653.

Cooper, G. (1996). "Theoretical Modelling and Biological Laws." *Philosophy of Science*, 63(3): S28–S35.

Couch, M. B. (2004). "Discussion—A Defense of Bechtel and Mundale." *Philosophy of Science*, 71 (2): 198–204.

Craver, C. (2004). "Dissociable Realization and Kind Splitting." *Philosophy of Science*, 71 (5): 960–971.

Craver, C. (2007). *Explaining the Brain: Mechanisms and the Mosaic Unity of Neuroscience*. Oxford: Oxford University Press.

Craver, C. (2008). "Physical Law and Mechanistic Explanation in the Hodgkin and Huxley Model of the Action Potential." *Philosophy of Science*, 75 (5): 1022–1033.

Craver, C. (2014). "The Ontic Account of Scientific Explanation." In M. I. Kaiser, O. R. Scholz, D. Plenge, and A. Hüttemann (Eds.), *Explanation in the Special Sciences: The Case of Biology and History*, 27–52. Dordrecht: Springer Netherlands.

Craver, C., and Bechtel, W. (2007). "Top-Down Causation without Top-Down Causes." *Biology & Philosophy*, 22 (4): 547–563.

Craver, C., and Darden, L. (2005). "Introduction". *Studies in History and Philosophy of Biological and Biomedical Science*, 36: 233–244.

Craver, C., and Tabery, J. (2016). "Mechanisms in Science." In E. N. Zalta (ed.), *The Stanford Encyclopedia of Philosophy* (Fall 2016 edition), http://plato.stanford.edu/archives/fall2016/entries/science-mechanisms/.

Crick, F. (1966). *Of Molecules and Men*. Seattle: University of Washington Press.

Cryder, C. E., Loewenstein, G., and Scheines, R. (2013). "The Donor is in the Details." *Organizational Behavior and Human Decision Processes*, 120 (1): 15–23.

da Costa, N. C. A., and French, S. (2003). *Science and Partial Truth: A Unitary Approach to Models and Scientific Reasoning*. Oxford: Oxford University Press.

Darrason, M. (2018). "Mechanistic and Topological Explanations in Medicine: The Case of Medical Genetics and Network Medicine." *Synthese*, 195 (1): 147–173.

Dawkins, R. (1986). *The Blind Watchmaker*. New York: Norton.

Delehanty, M. (2005). "Emergent Properties and the Context Objection to Reduction". *Biology and Philosophy*, 20 (4): 715–734.

DesAutels, L. (2010). "Sober and Elgin on Laws of Biology: A Critique." *Biology and Philosophy*, 25 (2): 249–256.

Dhar, P. K., and Giuliani, A. (2010). "Laws of Biology: Why so Few?" *Systems and Synthetic Biology*, 4 (1): 7–13.

Dorato, M. (2012). "Mathematical Biology and the Existence of Biological Laws." In D. Dieks, W. J. Gonzalez, S. Hartmann, M. Stöltzner, and M. Weber (Eds.), *Probabilities, Laws and Structure*, 109–121. Dordrecht: Springer.

Downes, S. M. (1992). "*The Importance of Models in Theorizing: A Deflationary Semantic View*." In *PSA: Proceedings of the Biennial Meeting of the Philosophy of Science Association*, 142–153. Chicago: The University of Chicago Press.

Dretske, F. (1977). "Laws of Nature". *Philosophy of Science*, 44: 248–268.

Dupré, J. (2013). "Living Causes'. *Proceedings of the Aristotelian Society Supplementary*, 87 (1): 19–38.

Earman, J. (1984). "Laws of Nature: The Empiricist Challenge". In R. J. Bogdan (Ed.), *D. M. Armstrong*, 191–223. Dordrecht: D. Reidel Publishing Company.

Earman, J. (2004). "Laws, Symmetry, and Symmetry Breaking: Invariance, Conservation Principles, and Objectivity." *Philosophy of Science*, 71 (5): 1227–1241.

Earman, J., and Roberts, J. (1999). "Ceteris Paribus, There Is No Problem of Provisos." *Synthese*, 118 (3): 439–478.

Earman, J., Roberts, J. T., and Smith, S. (2002). "Ceteris Paribus Lost." In J. Earman (Ed.), *Ceteris Paribus Laws*, 5–25. Dordrecht: Springer.

Elgin, M. (2003). "Biology and a Priori Laws". *Philosophy of Science*, 70 (5): 1380–1389.

Elgin, M. (2006). "There May Be Strict Empirical Laws in Biology, after All". *Biology and Philosophy*, 21 (1): 119–134.

Elgin, M., and Sober, E. (2002). "Cartwright on Explanation and Idealization." In J. Earman, C. Glymour, and S. Mitchell (Eds.), *Ceterus Paribus Laws*, 165–174. Dordrecht: Springer Netherlands.

Ellis, B. (2001). *Scientific Essentialism*. Cambridge: Cambridge University Press.

Ellis, B. (2002). *The Philosophy of Nature: A Guide to the New Essentialism*. Chesham, Buckinghamshire, UK: Acumen.

Ellis, B. (2010). *The Metaphysics of Scientific Realism*. Montreal: McGill-Queen's University Press.

Ellis, B., and Lierse, C. (1994). "Dispositional Essentialism". *Australasian Journal of Philosophy*, 72: 27–45.

Ereshefsky, M. (2009). "Natural Kinds in Biology." In *Routledge Encyclopedia of Philosophy*: https://www.rep.routledge.com/articles/natural-kinds-in-biology/v-1/.

Fales, E. (1990). *Causation and Universals*. London and New York: Routledge.

Fang, W. (2017). "Holistic Modeling: An Objection to Weisberg's Weighted Feature-Matching Account." *Synthese*, 194 (5): 1743–1764.

Fang, W. (2018). "The Case for Multiple Realization in Biology." *Biology & Philosophy* 33 (1): 3.

Fang, W.. (2019a). "Mixed-Effects Modeling and Nonreductive Explanation." *Philosophy of Science* 86 (5): 882–894.

Fang, W. (2019b). "An Inferential Account of Model Explanation." *Philosophia*, 47 (1): 99–116.

Fang, W. (2020a). "Multiple Realization in Systems Biology." *Philosophy of Science*, 87 (4): 663–684.

Fang, W. (2020b). "Multilevel Modeling and the Explanatory Autonomy of Psychology." *Philosophy of the Social Sciences*, 50 (3): 175–194.

Felline, L. (2018). "Mechanisms Meet Structural Explanation." *Synthese*, 195: 99–114.

Fisher, R. (1930). *The Genetic Theory of Natural Selection*. Oxford: Oxford University Press.

Fodor, J. (1974). "Special Sciences (or: The Disunity of Sciences as a Working Hypothesis)." *Synthese*, 28 (2): 97–115.

Fodor, J. (1991). "You Can Fool Some of the People All of the Time, Everything Else Being Equal; Hedged Laws and Psychological Explanations." *Mind*, 100 (397): 19–34.

French, S. (1997). "Partiality, Pursuit and Practice". In M. L. Dalla Chiara, K. Doets, D. Mundici, and J. van Bentham (Eds.), *Structures and Norms in Science*, 35–52. Dordrecht: Kluwer Academic Publishers.

French, S. (2003). "A Model-Theoretic Account of Representation (Or, I Don't Know Much about Art...but I Know It Involves Isomorphism)". *Philosophy of Science*, 70 (5): 1472–1483.

French, S., and Ladyman, J. (1998). "Semantic Perspective on Idealization in Quantum Mechanics". In N. Shanks (Ed.), *Idealization in Contemporary Physics. Poznan Studies in the Philosophy of the Sciences and the Humanities* (Vol. 63), 51–73. Amsterdam: Rodopi.

Frigg, R. (2006). "Scientific Representation and the Semantic View of Theories." *Theoria*, 21 (1): 49–65.

Frisch, M. F. (1998). *Theories, Models, and Explanation. Doctor dissertation.* Berkeley: University of California.

Frost-Arnold, G. (2004). "How to Be An Anti-Reductionist about Developmental Biology: Response to Laubichler and Wagner". *Biology and Philosophy*, 19 (1): 75–91.

Fung, D. C., Lo, A., Jankova, L., Clarke, S. J., Molloy, M., Robertson, G. R., and Wilkins, M. R. (2011). "Classification of Cancer Patients Using Pathway Analysis and Network Clustering." In G. Cagney, and A. Emili (Eds.), *Network Biology: Methods and Applications*, 311–336. Totowana, NJ: Humana Press.

Giere, R. (1988). *Explaining Science: A Cognitive Approach*. Chicago: University of Chicago Press.

Giere, R. (1999a). *Science without Laws*. Chicago: University of Chicago Press.

Giere, R. (1999b). "Using Models to Represent Reality." In L. Magnani, N. J. Nersessian, and P. Thagard (Eds.), *Model-based Reasoning in Scientific Discovery*, 41–57. Berlin: Springer Science & Business Media.

Giere, R. (2004). "How Models are Used to Represent Reality." *Philosophy of Science*, 71 (5): 742–752.

Giere, R. (2010). "An Agent-Based Conception of Models and Scientific Representation." *Synthese*, 172 (2): 269–281.

Gilbert, S. F., and Sarkar, S. (2000). "Embracing Complexity: Organicism for the 21st Century". *Developmental Dynamics*, 219 (1): 1–9.

Gillett, C. (2002). "The Dimensions of Realization: A Critique of the Standard View." *Analysis*, 62 (4): 316–323.

Gillett, C. (2003). "The Metaphysics of Realization, Multiple Realizability, and the Special Sciences.", *The Journal of Philosophy*, 100 (11): 591–603.

Gillett, C. (2007). "Understanding the New Reductionism: the Metaphysics of Science and Compositional Reduction." *The Journal of Philosophy*, 104 (4): 193–216.

Gillett, C. (2010). "Moving beyond the Subset Model of Realization: the Problem of Qualitative Distinctness in the Metaphysics of Science." *Synthese*, 177 (2): 165–192.

Ginzburg, L. R., and Colyvan, M. (2004). *Ecological Orbits: How Planets Move and Populations Grow*. Oxford: Oxford University Press.

Glennan, S. (2002). "Rethinking Mechanistic Explanation". *Philosophy of Science*, 69 (3): S342–353.

Glennan, S. (2007). "Mechanisms, Causes, and the Layered Model of the World". *Philosophy and Phenomenological Research*, 81 (2): 362–381.

Glennan, S. (2009). "Mechanisms". In H. Beebee, C. Hitchcock, and P. Menzies (Eds.), *The Oxford Handbook of Causation*. Oxford: Oxford University Press.

Glennan, S. (2010). "Mechanisms, Causes, and the Layered Model of the World." *Philosophy and Phenomenological Research* 81 (2): 362–381.

Godfrey-Smith, P. (2006). "The Strategy of Model-based Science." *Biology and Philosophy*, 21 (5): 725–740.

Goldenfeld, N. (1992). *Lectures on Phase Transitions and the Renormalization Group*. Boston: Addison-Wesley Publishing Company.

Goles, E., Schulz, O., and Markus, M. (2001). "Prime Number Selction of Cycles in a Predator-Prey Model". *Complexity*, 6 (4): 33–38.

Goodman, N. (1972). "Seven Strictures on Similarity." In N. Goodman (Ed.), *Problems and Projects*, 437–446. Indianapolis: Bobbs-Merril.

Gould, S. J. (1989). *Wonderful Life: The Burgess Shale and the Nature of History*. New York: W. W. Norton & Company.

Gouvêa, D. Y. (2015). "Explanation and the Evolutionary First Law(s)." *Philosophy of Science*, 82 (3): 363–382.

Greene, W. H. (2012). *Econometric Analysis* (7th edition). New York: Pearson.

Greenwood, P. E., and Nikulin, M. S. (1996). *A Guide to Chi-Squared Testing (Vol. 280)*. New York: John Wiley & Sons.

Griesemer, J. R. (1990). "Material Models in Biology". *Philosophy of Science*, 1990 (2): 79–93.

Griffiths, P. (1999). "Squaring the Circle: Natural Kinds with Historical Essences." In R. A. Wilson (Ed.), *Species: New Interdisciplinary Essays*, 209–228. Cambridge: MIT Press.

Gunning, P. W. (2001). "*Protein Isoforms and Isozymes*." In *Encyclopedia of Life Sciences*. John Wiley & Sons, Ltd. DOI: 10.1038/npg.els.0005717.

Gursoy, A., Keskin, O., and Nussinov, R. (2008). "Topological Properties of Protein Interaction Networks from a Structural Perspective." *Biochemical Society Transactions*, 36(Pt 6): 1398–1403.

Haas, L. W. (1977). "The Effect of the Spring-Neap Tidal Cycle on the Vertical Salinity Structure of the James, York and Rappahannock Rivers, Virginia, U.S.A". *Estuarine and Coastal Marine Science*, 5: 485–496.

Hamilton, W. D. (1967). "Extraordinary Sex Ratios". *Science*, 156 (3774): 477–488.

Hanahan, D., and Weinberg, R. A. (2000). "The Hallmarks of Cancer." *Cell*, 100 (1): 57–70.

Haufe, C. (2013). "From Necessary Chances to Biological Laws." *The British Journal for the Philosophy of Science*, 64 (2): 279–295.

Hausman, D. M. (1992). *The Inexact and Separate Science of Economics*. Cambridge: Cambridge University Press.

Hawkins, D. M. (2004). "The Problem of Overfitting." *Journal of Chemical Information and Computer Sciences*, 44 (1): 1–12.

Heaton, J. (2015). *Artificial Intelligence for Humans, Volume 3: Deep Learning and Neural Networks*. South Carolina: CreateSpace Independent Publishing Platform.

Heil, J. (1992). *The Nature of True Minds*. Cambridge: Cambridge University Press.

Heil, J. (1999). "Multiple Realizability." *American Philosophical Quarterly*, 36 (3): 189–208.

Hempel, C. (1965). *Aspects of Scientific Explanation and Other Essays in the Philosophy of Science*. New York: The Free Press.

Hempel, C., and Oppenheim, P. (1948). "Studies in the Logic of Explanation". *Philosophy of Science*, 15 (2): 135–175.

Hitchcock, C., and Woodward, J.. (2003). "Explanatory Generalizations, Part II: Plumbing Explanatory Depth". *Noûs*, 37 (2): 181–199.

Hopkins, B., and Skellam, J. G. (1954). "A New Method for Determining the Type of Distribution of Plant Individuals." *Annals of Botany*, 18(2): 213–227.

Huggins, E. M., and Schultz, E. A. (1967). "San Francisco Bay in A Warehouse". *Journal of the IEST*, 10 (5): 9–16.

Huggins, E. M., and Schultz, E. A. (1973). "The San Francisco Bay and the Delta Model". *California Engineer*, 51 (3): 11–23.

Hull, D. (1972). "Reductionism in Genetics—Biology or Philosophy?" *Philosophy of Science*, 39 (4): 491–499.

Hull, D. (1974). *Philosophy of Biological Science*. Englewood Cliffs, NJ: Prentice-Hall.

Hull, D. (1976). "Informal Aspects of Theory Reduction". In R.S. Cohen, and A. Michalos (Eds.), *Proceedings of the 1974 meeting of the Philosophy of Science Association*, 653–670. Dordrecht: D. Reidel.

Hull, D. (1979). "Discussion: Reduction in Genetics." *Philosophy of Science*, 46: 316–320.

Huneman, P.. (2010). "Topological Explanations and Robustness in Biological Sciences". *Synthese*, 171 (2): 213–245.

Huneman, P. (2018). "Diversifying the Picture of Explanations in Biological Sciences: Ways of Combining Topology with Mechanisms." *Synthese*, 195 (1): 115–146.

Hüttemann, A., and Love, A. C. (2011). "Aspects of Reductive Explanation in Biological Science: Intrinsicality, Fundamentality, and Temporality." *British Journal for the Philosophy of Science*, 62 (3): 519–549.

Illari, P., and Williamson, J. (2011). "Mechanisms Are Real and Local." In F. Russo and J. Williamson (Eds.), *Causality in the Sciences*, 818–844. Oxford: Oxford University Press.

Iynedjian, P. B. (1993). "Mammalian Glucokinase and Its Gene." *Biochemical Journal*, 293 (Pt1): 1–13.

Jackson, F., and Pettit, P. (1990). "Program Explanation: A General Perspective." *Analysis*, 50 (2): 107–117.

Jackson, F., and Pettit, P. (1992). "In Defense of Explanatory Ecumenism." *Economics and Philosophy*, 8 (1): 1–21.

Jackson, W. T., and Peterson, A. M. (1977). *The Sacramento-San Joaquin Delta: The evolution and Implementation of Water Policy*. Davis: California Water Resource Center, University of California.

Jenkins, C. S. 2008. "Romeo, René, and the Reasons Why: What Explanation Is." In *Proceedings of the Aristotelian Society (Hardback)*, 108: 61–84. Hoboken: Wiley Online Library.

Jöreskog, K. G. (1967). "Some Contributions to Maximum Likelihood Factor Analysis." *Psychometrika*, 32 (4): 443–482.

Jöreskog, K. G. (1969). "A General Approach to Confirmatory Maximum Likelihood Factor Analysis." *Psychometrika*, 34 (2): 183–202. doi: 10.1007/BF02289343.

Jöreskog, K. G. (1970a). "A General Method for Analysis of Covariance Structures." *Biometrika*, 57 (2): 239–251.

Jöreskog, K. G. (1970b). "A General Method for Estimating a Linear Structural Equation System." *ETS Research Bulletin Series*, 1970 (2): i–41.

Kaiser, M., and Hilgetag, C. C. (2004). "Spatial Growth of Real-World Networks." *Physical Review E*, 69 (3): 036103-1-036103-5.

Kaiser, M. I. (2015). *Reductive Explanation in the Biological Sciences*. Cham: Springer.

Kalckar, H. M. (1985). "The Discovery of Hexokinase." *Trends in Biochemical Sciences*, 10 (7): 291–293.

Kanno, H. (2000). "Hexokinase: Gene Structure and Mutations." *Best Practice & Research Clinical Haematology*, 13 (1): 83–88.

Kantarcı, B., and Labatut, V. (2013). "*Classification of Complex Networks Based on Topological Properties.*" *3rd Conference on Social Computing and its Applications*, Karlsruhe, Germany. doi:10.1109/CGC.2013.54.

Kauffman, S. A. (1976). "Articulation of Parts Explanation in Biology and the Rational Search for Them." In M. Grene, and E. Mendelsohn (Eds.), *Topics in the Philosophy of Biology*, 245–263. Dordrecht: Springer Netherlands.

Keesling, J. W. (1973). *Maximum Likelihood Approaches to Causal Flow Analysis*. Chicago: University of Chicago.

Keller, E. F. (1999). "Understanding Development". *Biology and Philosophy*, 14 (3): 321–330.

Kennedy, A. G. (2012). "A Non Representationalist View of Model Explanation." *Studies in History and Philosophy of Science Part A*, 43 (2): 326–332.

Kim, J. (1984). "Concepts of Supervenience." *Philosophy and Phenomenological Research*, 45 (2): 153–176.

Kim, J. (1992). "Multiple Realization and the Metaphysics of Reduction." *Philosophy and Phenomenological Research*, 52 (1): 1–26.

Kim, J. (1998). *Mind in a Physical World*. Cambridge: MIT Press.

Kim, J. (1999). "Making Sense of Emergency." *Philosophical Studies*, 95 (1/2): 3–36.

King, M. (2016). "On Structural Accounts of Model-Explanations." *Synthese*, 193 (9): 2761–2778.

Kistler, M. (2013). "The Interventionist Account of Causation and Non-Causal Association Laws." *Erkenntnis*, 78 (1): 65–84.

Kitcher, P. (1984). "1953 and All That. A Tale of Two Sciences." *The Philosophical Review*, 93 (3): 335–373.

Klein, C. (2008). "An Ideal Solution to Disputes about Multiply Realized Kinds." *Philosophical Studies*, 140 (2): 161–177.

Klein, C. (2013). "Multiple Realizability and the Semantic View of Theories." *Philosophical Studies*, 163 (3): 683–695.

Kleinbaum, D. G., and Klein, M. (2010). "Maximum Likelihood Techniques: An Overview". In D. G. Kleinbaum, and M. Klein (Eds.), *Logistic regression—A Self-Learning Text*, 103–127. New York: Springer New York.

Kostić, D. (2018). "The Topological Realization." *Synthese*, 195 (1): 79–98.

Kotz, S., Balakrishnan, N., and Johnson, N. L. (2000). *Continuous Multivariate Distributions, Volume 1, Models and Applications* (Vol. 59). New York: John Wiley & Sons.

Koyutürk, M., Kim, Y., Subramaniam, S., Szpankowski, W., and Grama, A. (2006). "Detecting Conserved Interaction Patterns in Biological Networks". *Journal of Computational Biology*, 13 (7): 1299–1322. doi:10.1089/cmb.2006.13.1299.

Kuhn, M., and Johnson, K. (2013). *Applied Predictive Modelling*. New York: Springer.

Lange, M. (1993). "Natural Laws and the Problem of Provisos." *Erkenntnis*, 38 (2): 233–248.

Lange, M. (2000). *Natural Laws in Scientific Practice*. Oxford: Oxford University Press.

Lange, M. (2002). "Who's Afraid of Ceteris-Paribus Laws? Or: How I Learned to Stop Worrying and Love Them." *Erkenntnis*, 57 (3): 407–423.

Lange, M. (2004). "The Autonomy of Functional Biology: A Reply to Rosenberg." *Biology and Philosophy*, 19 (1): 93–109.

Lange, M. (2005). "Ecological Laws: What Would They Be and Why Would They Matter?" *Oikos*, 110: 394–403.

Lange, M. (2007). "Laws and Meta-Laws of Nature: Conservation Laws and Symmetries." *Studies in History and Philosophy of Science Part B: Studies in History and Philosophy of Modern Physics*, 38 (3): 457–481.

Lange, M. (2009). *Laws and Lawmakers: Science, Metaphysics, and the Laws of Nature*. Oxford: Oxford University Press.

Lange, M. (2013). "What Makes a Scientific Explanation Distinctively Mathematical?" *British Journal for the Philosophy of Science*, 64 (3): 485–511.

LaPorte, J. (2004). *Natural Kinds and Conceptual Change*. Cambridge: Cambridge University Press.

Laubichler, M. D., and Wagner, G. P. (2001). "How Molecular is Molecular Developmental Biology? A Reply to Alex Rosenberg's Reductionism Redux: Computing the Embryo". *Biology and Philosophy*, 16 (1): 53–68.

Lawton, J. H. (1999). "Are There General Laws in Ecology?" *Oikos*, 84 (2): 177–192.

Lee, S., and Hershberger, S. (1990). "A Simple Rule for Generating Equivalent Models in Covariance Structure Modelling." *Multivariate Behavioral Research*, 25 (3): 313–334.

Levandowsky, M., and Winter, D. (1971). "Distance Between Sets." *Nature*, 234 (5): 34–35.

Levins, R. (1966). "The Strategy of Model Building in Population Biology." *American Scientist*, 54 (4): 421–431.

Lewis, D. (1973). *Counterfactuals*. Cambridge: Harvard University Press.

Lewis, D. (1983). "New Work for a Theory of Universals". *Australasian Journal of Philosophy*, 61: 343–377.

Lewis, D. (1986). *Philosophical Papers (Volume II)*. New York: Oxford University Press.

Lewis, D. (1994). "Humean Supervenience Debugged". *Mind*, 103: 473–490.

Lihoreau, M., Buhl, J., Charleston, M. A., Sword, G. A., Raubenheimer, D., and Simpson, S. J. (2014). "Modelling Nutrition across Organizational Levels: From Individuals to Superorganisms." *Journal of Insect Physiology*, 69: 2–11.

Liscaljet, I. M., Kleizen, B., and Braakman, I. (2005). "Studying Protein Folding in Vivo." In J. Buchner and T. Kiefhaber (Eds), *Protein Folding Handbook. Part II (Volume 1)*, 73–104. Weinheim: Wiley-VCH Verlag.

Lloyd, E. A. (1994). *The Structure and Confirmation of Evolutionary Theory*. Princeton: Princeton University Press.

Loewer, B. (1996). "Humean Supervenience". *Philosophical Topics*, 24: 101–126.

Longo, G., Montévil, M. R., and Pocheville, A. (2012). "From Bottom-up Approaches to Levels of Organization and Extended Critical Transitions." *Frontiers in Physiology*, 3: 232.

Lorenzano, P. (2006). "Fundamental Laws and Laws of Biology." In G. Ernst and K. G. Niebergall (Eds.), *Philosophie der Wissenschaft – Wissenschaft der Philosophie. Festschrift für C.Ulises Moulines zum 60. Geburstag* (Volume 5), 129–155. Mentis.

Love, A. (2008). "Explaining Evolutionary Innovations and Novelties: Criteria of Explanatory Adequacy and Epistemological Prerequisites". *Philosophy of Science*, 75 (5): 874–886.

Lyon, A. (2012). "Mathematical Explanations of Empirical Facts, and Mathematical Realism". *Australasian Journal of Philosophy*, 90 (3): 559–578.

Lyon, A., and Colyvan, M. (2008). "The Explanatory Power of Phase Spaces". *Philosophia Mathematica*, 16 (2): 227–243.

MacCallum, R. C., Wegener, D. T., Uchino, B. N., and Fabrigar, L. R. (1993). "The Problem of Equivalent Models in Applications of Covariance Structure Analysis." *Psychological Bulletin*, 114 (1): 185–199.

Machamer, P. (2004). "Activities and Causation: The Metaphysics and Epistemology of Mechanisms". *International Studies in the Philosophy of Science*, 18 (1): 27–39.

Machamer, P., Darden, L., and Craver, C. (2000). "Thinking About Mechanisms". *Philosophy of Science*, 67 (1): 1–25.

Mainx, F. (1955). "Foundations of Biology." *International Encyclopedia of Unified Science*, 1: 1–86.

Martindale, C. (1991). *Cognitive Psychology: A Neural-Network Approach*. Pacific Grove, CA: Thomson Brooks/Cole Publishing Co.

Masel, J., and Siegal, M. L. (2009). "Robustness: Mechanisms and Consequences." *Trends in Genetics: TIG*, 25 (9): 395–403.

Matthewson, J., and Weisberg, M. (2009). "The Structure of Tradeoffs in Model Building". *Synthese*, 170 (1): 169–190.

Mayr, E. (1969). "Discussion: Footnotes on the Philosophy of Biology." *Philosophy of Science*, 36: 197–202.

Mayr, E. (1988). *Toward a New Philosophy of Biology: Observations of an Evolutionist*. Cambridge: Harvard University Press.

Mayr, E. (1996). "The Autonomy of Biology: The Position of Biology Among the Sciences." *The Quarterly Review of Biology*, 71 (1): 97–106.

McMullin, E. (1978). "Structural Explanation." *American Philosophical Quarterly*, 15: 139–147.

McMullin, E. (1984). "A Case for Scientific Realism." In J. Leplin (Ed.), *Scientific Realism*, 8–40. Berkeley, CA: University of California Press.

McMullin, E. (1985). "Galilean Idealization." *Studies in History and Philosophy of Science Part A*, 16 (3): 247–273.

Miller, J. H., and Page, S. E. (2007). *Complex Adaptive Systems: An Introduction to Computational Models of Social Life*. Princeton: Princeton University Press.

Millikan, R. G. (1999). "Historical Kinds and the 'Special Sciences'." *Philosophical Studies*, 95 (1–2): 45–65.

Mitchell, S. (1997). "Pragmatic Laws". *Philosophy of Science*, 64 (4): S468–S479.

Mitchell, S. (2000). "Dimensions of Scientific Law". *Philosophy of Science*, 67 (2): 242–265.

Molnes, J. (2012). "Human Pancreatic Glucokinase: Structural and Physico-Chemical Studies related to Catalytic Activation, Kinetic Cooperativity and GCK-Diabetes." PhD diss., University of Bergen.

Montévil, M., and Pocheville, A. (2017). "The Hitchhiker's Guide to the Cancer Galaxy: How Two Critics Missed their Destination." *Organisms. Journal of Biological Sciences*, 1 (2): 37–48.

Montoya, J. M., Pimm, S. L., and Solé, R. V. (2006). Ecological Networks and Their Fragility." *Nature*, 442 (7100): 259–264.

Morgan, G. J. (2010). "Laws of Biological Design: A Reply to John Beatty." *Biology and Philosophy*, 25 (3): 379–389.

Morgan, M. S., and Morrison, M. (1999). *Models as Mediators: Perspectives on Natural and Social Science (Vol. 52)*. Cambridge: Cambridge University Press.

Morrison, M. (1999). "Models as Autonomous Agents". In M. Morgan, and M. Morrison (Eds.), *Models as Mediators: Perspectives on Natural and Social Science*, 38–65. Cambridge: Cambridge University Press.

Mostafavi, S., Goldenberg, A., and Morris, Q. (2011). "Predicting Node Characteristics from Molecular Networks." In G. Cagney, and A. Emili (Eds.), *Network Biology: Methods and Applications*, 399–414. Totowa: Humana Press.

Müller, G. B., and Wagner, G. P. (1996). "Homology, Hox Genes, and Developmental Integration". *American Zoologist*, 36 (1): 4–13.

Munson, R. (1975). "Is Biology a Provincial Science?" *Philosophy of Science*, 42:428–447.

Murray, B. G. (1992). "Research Methods in Physics and Biology." *Oikos*, 64 (3): 594–596.

Murray, B. G. (1999). "Is Theoretical Ecology A Science? A Reply to Turchin (1999)." *Oikos*, 87 (3): 594–600.

Murray, B. G. (2000). "Universal Laws and Predictive Theory in Ecology and Evolution." *Oikos*, 89: 403–408.

Musgrave, A. (1981). "'Unreal Assumptions' in Economic Theories: The F-Twist Untwist." *Kyklos*, 34 (3): 377–387.

Myung, I. J. (2003). "Tutorial on Maximum Likelihood Estimation." *Journal of Mathematical Psychology*, 47 (1): 90–100.

Nagel, E. (1961). *The Structure of Science: Problems in the Logic of Scientific Explanation*. San Diego: Harcourt, Brace & World.

Nelson, D. L., and Cox, M. M. (2008). *Lehninger Principles of Biochemistry* (4th edition). New York: W. H. Freeman and Company.

Nievergelt, Y. (2000). "A Tutorial History of Least Squares with Applications to Astronomy and Geodesy." *Journal of Computational and Applied Mathematics*, 121 (1–2): 37–72.

Nowak, M. A. (2006). "Five Rules for the Evolution of Cooperation". *Science*, 314 (5805): 1560–1563.

O'Hara, R. B. (2005). "The Anarchist's Guide to Ecological Theory: Or, We Don't Need no Stinkin' Laws." *Oikos*, 110: 390–393.

Odenbaugh, J. (2003). "Complex Systems, Trade-Offs, and Theoretical Population Biology: Richard Levins's 'Strategy of Model Building in Population Biology' Revisited." *Philosophy of Science*, 70 (5): 1496–1507.

Odenbaugh, J. (2006). "The Strategy of 'The strategy of Model Building in Population Biology'." *Biology and Philosophy*, 21 (5): 607–621.

Odenbaugh, J. (2008). "Models." In S. Sarkar, and A. Plutynski (Eds.), *A Companion to the Philosophy of Biology*, 506–524. Hoboken: Blackwell Publishing.

Odenbaugh, J. (2014). "Semblance or Similarity? Reflections on Simulation and Similarity". *Biology and Philosophy*, 30 (2): 277–291.

Okasha, S. (2002). "Darwinian Metaphysics: Species and the Question of Essentialism." *Synthese*, 131 (2): 191–213.

Orilia, F., and Swoyer, C. (2016). "Properties." In Edward N. Zalta (Ed.), *The Stanford Encyclopedia of Philosophy* (Winter 2016 edition), https://plato.stanford.edu/archives/win2016/entries/properties/.

Orzack, S. H. (2005). "What, If Anything, Is 'The Strategy of Model Building in Population Biology?' A Comment on Levins (1966) and Odenbaugh (2003)." *Philosophy of Science*, 72 (3): 479–485.

Orzack, S. H., and Sober, E. (1993). "A Critical Assessment of Levins's The Strategy of Model Building in Population Biology (1966)." *The Quarterly Review of Biology*, 68 (4): 533–546.

Otsuka, J. (2021) "When Less Is More: A Statistical Look at Level of Explanation". In T. Matsuda, J. Wolff, and T. Yanagawa (Eds.), *Risks and Regulation of New Technologies*, 47–65. Singapore: Springer.

Parker, W. S. (2009). "Confirmation and Adequacy-for-Purpose in Climate Modelling." *Proceedings of the Aristotelian Society*, Supplementary Volumes, 83: 233–249.

Parker, W. S. (2010). "Scientific Models and Adequacy-for-Purpose." *Modern Schoolman: A Quarterly Journal of Philosophy (Proceedings of the 2010 Henle Conference on Experimental & Theoretical Knowledge)*, 87(3–4): 285–293.

Parker, W. S. (2015). "Getting (even more) Serious about Similarity." *Biology and Philosophy*, 30 (2): 267–276.

Peregrín-Alvarez, J. M., Sanford, C., and Parkinson, J. (2009) "The Conservation and Evolutionary Modularity of Metabolism." *Genome Biology*, 10 (6): 1–17.

Pexton, M. (2014). "How Dimensional Analysis Can Explain." *Synthese*, 191 (10): 2333–2351.

Piccinini, G., and Maley, C. J. (2014). "The Metaphysics of Mind and the Multiple Sources of Multiple Realizability." In M. Sprevak, and J. Kallestrup (Eds.), *New Waves in Philosophy of Mind*, 125–152. London: Palgrave Macmillan.

Pietroski, P., and Rey, G. (1995). "When Other Things Aren't Equal: Saving Ceteris Paribus Laws from Vacuity." *The British Journal for the Philosophy of Science*, 46 (1): 81–110.

Pincock, C. (2005). "Overextending Partial Structures: Idealization and Abstraction". *Philosophy of Science*, 72 (5): 1248–1259.

Pincock, C. (2007). "A Role for Mathematics in the Physical Sciences". *Noûs*, 41 (2): 253–275.

Pincock, C. (2011). *Mathematics and Scientific Representation*. Oxford: Oxford University Press.

Pincock, C. (2015). "Abstract Explanations in Science". *The British Journal for the Philosophy of Science*, 66 (4): 857–882.

Pincock, C. (2018). "Accommodating Explanatory Pluralism". In A. Reutlinger, and J. Saatsi (Eds.), *Explanation Beyond Causation*. Oxford: Oxford University Press.

Pocheville, A., Griffiths, P. E., and Stotz, K. 2017. "Comparing Causes – an Information-Theoretic Approach to Specificity, Proportionality and Stability." In *Proceedings of the 15th Congress of Logic, Methodology and Philosophy of Science*. London: College Publications.

Polger, T. W. (2008). "Two Confusions Concerning Multiple Realization." *Philosophy of Science*, 75 (5): 537–547.

Polger, T. W. (2009). "Evaluating the Evidence for Multiple Realization." *Synthese*, 167 (3): 457–472.

Polger, T. W., and Shapiro, L. A. (2016). *The Multiple Realization Book*. Oxford: Oxford University Press.

Pollard, D., and Radchenko, P. (2006). "Nonlinear Least-Squares Estimation." *Journal of Multivariate Analysis*, 97 (2): 548–562.

Potochnik, A. (2017). *Idealization and the Aims of Science*. Chicago: University of Chicago Press.

Press, J. (2009). "Physical Explanations and Biological Explanations, Empirical Laws and a Priori Laws." *Biology and Philosophy*, 24 (3): 359–374.

Press, W. H., Teukolsky, S. A., Vetterling, W. T., and Flannery, B. P. (1992). *Numerical Recipes in C: The Art of Scientific Computing* (2nd edition). Cambridge: Cambridge University Press.

Putnam, H. (1967). "Psychological Predicates." In W. H. Capitan, and D. D. Merrill (Eds.), *Art, Mind, and Religion*, 37–48. Pittsburgh: University of Pittsburgh Press.

Putnam, H. (1975). "Philosophy and our Mental Life". In *Mind, Language, and Reality*, 291–303. Cambridge: Cambridge University Press.

Quenette, P. Y., and Gerard, J. F. (1993). "Why Biologists Do Not Think like Newtonian Physicists." *Oikos*, 68 (2): 361–363.

Quine, W. V. O. (1969). "Natural Kinds." In W. V. O. Quine (Ed.), *Ontological Relativity and Other Essays*, 114–138. New York: Columbia University Press.

Raerinne, J. (2011). "Causal and Mechanistic Explanations in Ecology." *Acta Biotheoretica*, 59 (3): 251–271.

Railton, P. (1981). "Probability, Explanation, and Information." *Synthese*, 48 (2): 233–256.

Ramsey, F. (1978/1928). *Foundations*. London: Routledge and Kegan Paul.

Rathkopf, C. (2018). "Network Representation and Complex Systems." *Synthese* 195 (1): 55–78.

Raykov, T., and Marcoulides, G. A. (2001). "Can There Be Infinitely Many Models Equivalent to a given Covariance Structure Model?" *Structural Equation Modelling*, 8 (1): 142–149.

Reutlinger, A. (2016). "Is There A Monist Theory of Causal and Non-causal Explanations? The Counterfactual Theory of Scientific Explanation." *Philosophy of Science*, 83 (5): 733–745.

Reutlinger, A. (2017). "Explanation beyond Causation? New Directions in the Philosophy of Scientific Explanation." *Philosophy Compass* 12 (2): e12395.

Reutlinger, A., Schurz, G., and Hüttemann, A. (2015). "Ceteris Paribus Laws." In E. N. Zalta (Ed.), *The Stanford Encyclopedia of Philosophy* (Fall 2015 edition), http://plato.stanford.edu/archives/fall2015/entries/ceteris-paribus/.

Rice, C. (2012). "Optimality Explanations: A Plea for An Alternative Approach." *Biology & Philosophy*, 27 (5): 685–703.

Rice, C. (2015). "Moving beyond Causes: Optimality Models and Scientific Explanation." *Noûs*, 49 (3): 589–615.

Rice, J. A. (2006). *Mathematical Statistics and Data Analysis* (3rd edition). Toronto: Nelson Education.

Richardson, R. C. (2008). "Autonomy and Multiple Realization." *Philosophy of Science*, 75 (5): 526–536.

Rieppel, O. (2010). "New Essentialism in Biology." *Philosophy of Science*, 77 (5): 662–673.

Robert, J. S. (2004). *Embryology, Epigenesis, and Evolution: Taking Development Seriously*. New York: Cambridge University Press.

Roe, A. W., Pallas, S. L., Hahm, J.-O., and Sur, M. (1990). "A Map of Visual Space Induced in Primary Auditory Cortex." *Science*, 250 (4982): 818–820.

Rohwer, Y., and Rice, C. (2016). "How Are Models and Explanations Related?" *Erkenntnis*, 81 (5): 1127–1148.

Rosenberg, A. (1985). *The Structure of Biological Science*. Cambridge: Cambridge University Press.

Rosenberg, A. (1997). "Reductionism Redux: Computing the Embryo". *Biology and Philosophy*, 12 (4): 445–470.

Rosenberg, A. (2001a). "Reductionism in a Historical Science". *Philosophy of Science*, 68 (2):135–163.

Rosenberg, A. (2001b). "How is Biological Explanation Possible?" *British Journal for the Philosophy of Science*, 52 (4):735–760.

Rosenberg, A. (2001c). "On Multiple Realization and the Special Sciences." *The Journal of Philosophy*, 98 (7): 365–373.

Rosenberg, A. (2006). *Darwinian Reductionism, or How to Stop Worrying and Love Molecular Biology*. Chicago: University of Chicago Press.

Rosenberg, A. (2012). "Why Do Spatiotemporally Restricted Regularities Explain in the Social Sciences?" *British Journal for the Philosophy of Science*, 63 (1): 1–26.

Rosenberg, A., and McShea, D. (2008). *Philosophy of Biology: A Contemporary Introduction*. Milton Park: Routledge.

Ruse, M. (1971). "Reduction, Replacement, and Molecular Biology." *Dialectica*, 25: 39–72.

Ruse, M. (1973). *The Philosophy of Biology*. London: Hutchinson.

Ruse, M. (1976). "Reduction in Genetics". In R.S. Cohen, and A. Michalos (Eds.), *Proceedings of the 1974 Meeting of the Philosophy of Science Association*, 633–651. Dordrecht: D. Reidel.

Ruthazer, E. S., and Stryker, M. P. (1996). "The Role of Activity in the Development of Long-Range Horizontal Connections in Area 17 of the Ferret." *The Journal of Neuroscience*, 16 (22): 7253–7269.

Saatsi, J., and Pexton, M. (2013). "Reassessing Woodward's Account of Explanation: Regularities, Counterfactuals, and Noncausal Explanations." *Philosophy of Science*, 80 (5): 613–624.

Sachs, J. L., Mueller, U. G., Wilcox, T. P., and Bull., J. J. (2004). "The Evolution of Cooperation." *The Quarterly Review of Biology*, 79 (2): 135–160.

Sachse, C. (2012). "Biological Laws and Kinds Within a Conservative Reductionist Framework." In D. Dieks, J. W. Gonzalez, S. Hartmann, M. Stöltzner, and M. Weber (Eds.), *Probabilities, Laws, and Structures*, 231–244. Dordrecht: Springer Netherlands.

Sagoff, M.. (2016). "Are There General Causal Forces in Ecology?" *Synthese*, 193 (9): 3003–3024.

Salmon, W. (1971). *Statistical Explanation and Statistical Relevance*. Pittsburgh: University of Pittsburgh Press.

Salmon, W. (1977). "A Third Dogma of Empiricism." In R. E. Butts and J. Hintikka (Eds.), *Basic Problems in Methodology and Linguistics*, 149–166. Dordrecht: Reidel.

Salmon, W. (1984). *Scientific Explanation and Causal Structure of the World*. Princeton: Princeton University Press.

Salmon, W. (1989). *Four Decades of Scientific Explanation*. Minneapolis: University of Minnesota Press.

Saramäki, J., Kivelä, M., Onnela, J.-P., Kaski, K., and Kertész, J. (2007). "Generalizations of the Clustering Coefficient to Weighted Complex Networks." *Physical Review E*, 75 (2): 027105-1-027108-4.

Sarkar, S. (1992). "Models of Reduction and Categories of Reductionism". *Synthese*, 91 (3): 167–194.

Sarkar, S. (1998). *Genetics and Reductionism*. Cambridge: Cambridge University Press.

Sarkar, S. (2001). "Reduction: A Philosophical Analysis". In *Encyclopedia of Life Sciences*, Hoboken, NJ: John Wiley & Sons, Ltd.

Sarkar, S. (2002). "Genes versus Molecules: How To, and How Not To, Be a Reductionist". In M. H. V. Van Regenmortel, and D. L. Hull (Eds.), *Promises and Limits of Reductionism in the Biomedical Sciences*. John Wiley & Sons, Ltd.

Sarkar, S. (2005). "Reductionism and Functional Explanation in Molecular Biology". In S. Sarkar (Ed.), *Molecular Models of Life: Philosophical Papers on Molecular Biology*. Cambridge, MA: MIT Press.

Savenije, H. H. G. (2005). "*Salinity and Tides in Alluvial Estuaries*". Amsterdam: Elsevier Science.

Schaffner, K. F. (1967). "Approaches to Reduction". *Philosophy of Science*, 34 (2): 137–147.

Schaffner, K. F. (1969). "The Watson-Crick Model and Reductionism". *British Journal for the Philosophy of Science*, 20 (4): 325–348.

Schaffner, K. F. (1974). "The Peripherality of Reductionism in the Development of Molecular Biology". *Journal of the History of Biology*, 7 (1): 111–139.

Schaffner, K. F. (1976). "Reductionism in Biology: Prospects and Problems". In R.S. Cohen, and A. Michalos (Eds.), *Proceedings of the 1974 Meeting of the Philosophy of Science Association*, 613–632. Dordrecht: D. Reidel.

Schaffner, K. F. (1993). *Discovery and Explanation in Biology and Medicine*. Chicago: Chicago University Press.

Schaffner, K. F. (1996). "Theory Structure and Knowledge Representation in Molecular Biology". In S. Sarkar (Ed.), *The Philosophy and History of Molecular Biology: New perspectives*, 27–46. Dordrecht: Kluwer.

Schaffner, K. F. (2006) "Reduction: The Cheshire Cat Problem and a Return to Roots." *Synthese*, 151 (3): 37–402.

Schiffer, S. (1991). "Ceteris Paribus Laws." *Mind*, 100 (1): 1–17.

Scriven, M. (1962). "Explanations, Predictions, and Laws". In H. Feigl, and G. Maxwell (Eds.), *Scientific Explanation, Space, and Time (Minnesota Studies in the Philosophy of Science: Vol. 3)*, 170–230. Minneapolis: University of Minnesota Press.

Senior, A. M., Charleston, M. A., Lihoreau, M., Buhl, J., Raubenheimer, D., and Simpson, S.J. (2015). "Evolving Nutritional Strategies in the Presence of Competition: A Geometric Agent-Based Model." *PLoS Comput Biol*, 11 (3): e1004111.

Şerban, M. (2017). "What Can Polysemy Tell Us about Theories of Explanation?" *European Journal for Philosophy of Science*, 7 (1): 41–56.

Shapiro, L. A. (2000). "Multiple Realizations." *The Journal of Philosophy*, 97 (12): 635–654.

Shapiro, L. A. (2004). *The Mind Incarnate*. Cambridge: MIT Press.

Shapiro, L. A. (2008). "How to Test for Multiple Realization." *Philosophy of Science*, 75 (5): 514–525.

Shapiro, L. A., and Polger, T. W. (2012). "Identity, Variability, and Multiple Realization in the Special Sciences." In S. Gozzano, and C. S. Hill (Eds.), *New Perspectives on Type Identity*, 264–287. Cambridge: Cambridge University Press.

Sharan, R., Suthram, S., Kelley, R. M., Kuhn, T., McCuine, S., Uetz, P., Sittler, T., Karp, R. M., and Ideker, T. (2005). "Conserved Patterns of Protein Interaction in Multiple Species." *Proceedings of the National Academy of Sciences of the United States of America*, 102 (6): 1974–1979.

Sharma, J., Angelucci, A., and Sur, M. (2000). "Induction of Visual Orientation Modules in Auditory Cortex." *Nature*, 404 (6780): 841–847.

Sheredos, B. (2016). "Re-Reconciling the Epistemic and Ontic Views of Explanation (or, Why the Ontic View Cannot Support Norms of Generality)." *Erkenntnis* 81 (5): 919–949.

Shipley, B. (2002). *Cause and Correlation in Biology: A User's Guide to Path Analysis, Structural Equations and Causal Inference*. Cambridge: Cambridge University Press.

Shipley, B., and Lechowicz, M. (2000). "The Functional Coordination of Leaf Morphology, Nitrogen Concentration, and Gas Exchange in 40 Wetland Plant Species". *Ecoscience*, 7 (2): 183–194.

Shoemaker, S. (1980). "Causality and properties." In P. van Inwagen (Ed.), *Time and Cause*, 109–135. Dordrecht: Reidel.

Shoemaker, S. (2001). "Realization and Mental Causation." In C. Gillett, and B. Loewer (Eds.), *Physicalism and its Discontents*. Cambridge: Cambridge University Press.

Simpson, G. G. (1964). *This View of Life*. New York: Harcourt, Brace & World.

Skyrms, B. (1980). *Causal Necessity*. New Haven: Yale University Press.

Skyrms, B., and Lambert, K. (1995). "The Middle Ground: Resiliency and Laws in the Web of Belief." In F. Weinert (Ed.), *Laws of Nature: Essays on the Philosophical, Scientific, and Historical Dimensions*, 139–156. Berlin: Walter de Gruyter.

Smart, J. J. C. (1963). *Philosophy and Scientific Realism*. London: Routledge & Kegan Paul.

Smith, J. M., and Price, G. R. (1973). "The Logic of Animal Conflict." *Nature*, 246 (2): 15–18.

Smith, T. A. D. (2000). "Mammalian Hexokinases and Their Abnormal Expression in Cancer." *British Journal of Biomedical Science*, 57 (2): 170–178.

Sneed, J. D. (1971). *The Logical Structure of Mathematical physics*. Dordrecht: Reidel.

Sober, E. (1983). 'Equilibrium Explanation'. *Philosophical Studies*, 43 (2): 201–210.

Sober, E. (1984). *The Nature of Selection: Evolutionary Theory in Philosophical Focus* (Vol. 95). Chicago: University of Chicago Press.

Sober, E. (1997). "Two Outbreaks of Lawlessness in Recent Philosophy of Biology". *Philosophy of Science*, 64 (4): S458–S467.

Sober, E. (1999). "The Multiple Realizability Argument against Reductionism." *Philosophy of Science*, 66 (4): 542–564.

Sober, E. (2000). *Philosophy of Biology* (Vol. 45). Colorado: Westview Press.

Sober, E. (2008). "Fodor's *Bubbe Meise* Against Darwinism." *Mind and Language*, 23 (1): 42–49.

Sober, E. (2010). "Natural Selection, Causality, and Laws: What Fodor and Piatelli-Palmarini Got Wrong." *Philosophy of Science*, 77 (4): 594–607.

Soffer, S. N., and Vázquez, A. (2005). "Network Clustering Coefficient without Degree-Correlation Biases." *Physical Review E*, 71 (5): 057101-1-057101-4.

Solnica-Krezel, L. (2005) "Conserved Patterns of Cell Movements during Vertebrate Gastrulation." *Current Biology*, 15 (6): R213–R228. doi:10.1016/j.cub.2005.03.016.

Sonnenschein, C., and Soto, A. M. (1999). *The Society of Cells: Cancer and Control of Cell Proliferation*. Milton Park: Bios Scientific Pub Limited.

Soto, A. M., and Sonnenschein, C. (2014). "One Hundred Years of Somatic Mutation Theory of Carcinogenesis: Is It Time to Switch? *BioEssays: News and Reviews in Molecular, Cellular and Developmental Biology*, 36 (1): 118–120.

Stegmüller, W. (1976). *The Structure and Dynamics of Theories*. New York: Springer-Verlag.

Steiner, M. (1978). "Mathematics, Explanation, and Scientific Knowledge". *Noûs*, 12 (1): 17–28.

Stelzl, I. (1986). "Changing a Causal Hypothesis without Changing the Fit: Some Rules for Generating Equivalent Path Models." *Multivariate Behavioral Research*, 21 (3): 309–331.

Sterelny, K., and Griffiths, P. (1999). *Sex and Death: An Introduction to Philosophy of Biology*. Chicago: University of Chicago Press.

Stotz, K. (2006). "Molecular Epigenesis: Distributed Specificity as a Break in the Central Dogma". *History and Philosophy of the Life Sciences*, 28 (4): 533–548.

Strevens, M. (2008). *Depth: An Account of Scientific Explanation*. Cambridge: Harvard University Press.

Suárez, M. (2003). "Scientific Representation: against Similarity and Isomorphism." *International Studies in the Philosophy of Science*, 17 (3): 225–244.

Suárez, M. (2004). "An Inferential Conception of Scientific Representation". *Philosophy of Science*, 71 (5): 767–779.

Suárez, M. (2015). "Deflationary Representation, Inference, and Practice". *Studies in History and Philosophy of Science*, 49 (2015): 36–47.

Suárez, M. (2016). "Representation in Science". In P. Humphreys (Ed.), *The Oxford Handbook of Philosophy of Science*. Oxford: Oxford University Press.

Suárez, M., and Cartwright, N. (2008). "Theories: Tools versus Models". *Studies in History and Philosophy of Modern Physics*, 39 (2008): 62–81.

Sullivan, J. A. (2008). "Memory Consolidation, Multiple Realizations, and Modest Reductions." *Philosophy of Science*, 75 (5): 501–513.

Suppe, F. (1974). "Theories and Phenomena." In W. Leinfellnerand, and E. Kohler (Eds.), *Developments of the Methodology of Social Science*, 45–92. Dordrecht: Reidel.

Suppe, F. (1977). *The Structure of Scientific Theories*. Urbana, Illinois: University of Illinois Press.

Suppe, F. (1989). *The Semantic Conception of Theories and Scientific Realism*. Urbana: University of Illinois Press.

Suppes, P. (1957). *Introduction to Logic*. New Jersey: D. Van Nostrand and Co.
Suppes, P. (1960). "A Comparison of the Meaning and Uses of Models in Mathematics and the Empirical Sciences." *Synthese*, 12 (2–3): 287–301.
Suppes, P. (1962). "Models of Data." In E. Nagel, P. Suppes, and A. Tarski (Eds.), *Logic, Methodology, and the Philosophy of Science*, 252–261. California: Stanford University Press.
Suppes, P. (1967). "What is a Scientific Theory?" In S. Morgenbesser (Ed.), *Philosophy of Science Today*. New York: Meridian Books.
Swoyer, C. (1982). "The Nature of Natural Laws". *Australasian Journal of Philosophy*, 60 (3): 203–223.
Symonds, M. R. E., and Blomberg, S. P. (2014). "A Primer on Phylogenetic Generalised Least Squares." In L. Z. Garamszegi (Ed.), *Modern Phylogenetic Comparative Methods and Their Application in Evolutionary Biology: Concepts and Practice*, 105–130. Berlin, Heidelberg: Springer Berlin Heidelberg.
Târziu, G. (2018). "Importance and Explanatory Relevance: The Case of Mathematical Explanations." *Journal for General Philosophy of Science* 49 (3): 393–412.
Taylor, T. (2000). "Socio-Ecological Webs and Sites of Sociality: Levins' Strategy of Model Building Revisited." *Biology and Philosophy*, 15 (2): 197–210.
Thompson, P. (1983). "The Structure of Evolutionary Theory: A Semantic Perspective". *Studies in History and Philosophy of Science*, 14 (3):
215–229.
Tooley, M. (1977). "The Nature of Laws". *Canadian Journal of Philosophy*, 7: 667–698.
Tooley, M. (1987). *Causation*. Oxford: Clarendon Press.
Turchin, P. (2001). "Does Population Ecology Have General Laws?" *Oikos*, 94: 17–26.
Usher, M., and McClelland, J. L. (2001). "The Time Course of Perceptual Choice: The Leaky, Competing Accumulator Model." *Psychological Review*, 108 (3): 550–592.
Utts, J. M., and Heckard, R. F. (2011). *Mind on Statistics* (4th edition). Boston: Cengage Learning.
van Baalen, M., and Rand, D. A. (1998). 'The Unit of Selection in Viscous Populations and the Evolution of Altruism'. *Journal of Theoretical Biology*, 193 (4): 631–648.
van Fraassen, B. C. (1970). "On the Extension of Beth's Semantics of Physical Theories." *Philosophy of Science*, 37 (3): 325–339.
van Fraassen, B. C. (1972). "A Formal Approach to the Philosophy of Science." In R. Colodny (Ed.), *Paradigms and Paradoxes*. Pittsburgh: University of Pittsburgh Press.
van Fraassen, B. C. (1974). "The Labyrinth of Quantum Logic." In R. S. Cohen, and M. Wartofsky (Eds.), *Logical and Empirical Studies in Contemporary Physics*. Dordrecht: D. Reidel Publishing Company.
van Fraassen, B. C. (1980). *The Scientific Image*. New York: Oxford University Press.
Von Melchner, L., Pallas, S. L., and Sur, M. (2000). "Visual Behaviour Mediated by Retinal Projections Directed to the Auditory Pathway." *Nature*, 404 (6780): 871–876.
Waskan, J. A. (2006). *Models and Cognition*. Cambridge: MIT Press.

Waters, C. K. (1990). "Why the Antireductionist Consensus Won't Survive the Case of Classical Mendelian Genetics". In A. Fine, M. Forbes, and L. Wessels (Eds.), *Proceedings of the biennial meeting of the Philosophy of Science Association* (Vol. 1), 125–139. East Lansing: University of Chicago Press.

Waters, C. K. (1994). "Genes Made Molecular". *Philosophy of Science*, 61 (2):163–185.

Waters, C. K. (2000). "Molecules Made Biological". *Revue Internationale de Philosophie*, 54 (214 (4)): 539–564.

Waters, C. K. (2008). "Beyond Theoretical Reduction and Layer-Cake Antireduction: How DNA Retooled Genetics and Transformed Biological Practice". In M. Ruse (Ed.), *The Oxford Handbook of Philosophy of Biology*, 238–262. New York: Oxford University Press.

Watts, D. J., and Strogatz, S. H. (1998). "Collective Dynamics of 'small-World'networks." *Nature*, 393 (6684): 440–442.

Weber, M. (2005). *Philosophy of Experimental Biology*. Cambridge: Cambridge University Press.

Weisberg, M. (2006). "Forty Years of 'The Strategy': Levins on Model Building and Idealization." *Biology and Philosophy*, 21 (5): 623–645.

Weisberg, M. (2012). "Getting Serious about Similarity." *Philosophy of Science*, 79 (5): 785–794.

Weisberg, M. (2013). *Simulation and Similarity: Using Models to Understanding the World*. New York: Oxford University Press.

Weisberg, M. (2015). "Biology and Philosophy symposium on Simulation and Similarity: Using Models to Understand the World." *Biology and Philosophy*, 30 (2): 299–310.

Weslake, B. (2010). "Explanatory Depth". *Philosophy of Science*, 77 (2): 273–294.

West, S. A., Griffin, A. S., and Gardner, A. (2007). "Social Semantics: Altruism, Cooperation, Mutualism, Strong Reciprocity and Group Selection." *Journal of Evolutionary Biology*, 20 (2): 415–432.

Wilensky, U., and Rand, W.. (2015). *An Introduction to Agent-Based Modelling: Modelling Natural, Social, and Engineered Complex Systems with NetLogo*. Cambridge: The MIT Press.

Wilke, C. O., Wang, J. L., Ofria, C., Lenski, R. E., and Adami, C. (2001). "Evolution of Digital Organisms at High Mutation Rates Leads to Survival of the Flattest." *Nature*, 412 (6844): 331–333.

Wilson, R. A. (2001). "Two Views of Realization." *Philosophical Studies*, 104 (1): 1–31.

Wimsatt, W. C. (1974). "Reductive Explanation: A Functional Account." *Philosophy of Science*, 1974 (Proceedings): S671–S710.

Wimsatt, W. C. (1976). "Reductionism, Levels of Organization, and the Mind-Body Problem." In G. G. Globus, G. Maxwell, and I. Savodnik (Eds.), *Consciousness and the Brain: A Scientific and Philosophical Inquiry*, 202–267. New York: Plenum Press.

Wimsatt, W. C. (1979). "Reductionism and Reduction". In P. D. Asquith, and H. E. Kyburg (Eds.), *Current Research in Philosophy of Science*, 352–377. East Lansing: Philosophy of Science Association.

Wimsatt, W. C. (1994). "The Ontology of Complex Systems: Levels of Organization, Perspectives, and Causal Thickets." *Canadian Journal of Philosophy*, 24 (supplement 1): 207–274.

Wimsatt, W. C. (2007). *Re-Engineering Philosophy for Limited Beings: Piecewise Approximations to Reality*. Cambridge: Harvard University Press.

Woodward, J. (1997). "Explanation, Invariance and Intervention". *Philosophy of Science*, 64: S26–S41.

Woodward, J. (2000). "Explanation and Invariance in the Special Sciences". *British Journal for the Philosophy of Science*, 51 (2): 197–254.

Woodward, J. (2001). "Law and Explanation in Biology: Invariance is the Kind of Stability That Matters". *Philosophy of Science*, 68 (1): 1–20.

Woodward, J. (2002). "There Is No Such Thing as a Ceteris Paribus Law." *Erkenntnis*, 57 (3): 303–328.

Woodward, J. (2003). *Making Things Happen: A Theory of Causal Explanation*. Oxford: Oxford University Press.

Woodward, J. (2010). "Causation in Biology: Stability, Specificity, and the Choice of Levels of Explanation". *Biology and Philosophy*, 25 (3): 287–318.

Woodward, J. (2013a). "Causation and Manipulability". In E. N. Zalta (Ed.)*The Stanford Encyclopedia of Philosophy* (Winter 2013 edition), http://plato.stanford.edu/archives/win2013/entries/causation-mani/.

Woodward, J. (2013b). "Mechanistic Explanation: Its Scope and Limits." *Proceedings of the Aristotelian Society Supplementary*, 87 (1): 39–65.

Woodward, J. (2014). "Scientific Explanation". *The Stanford Encyclopedia of Philosophy* http://plato.stanford.edu/archives/win2014/entries/scientific-explanation/.

Woodward, J., and Hitchcock, C. (2003). "Explanatory Generalizations, Part I: A Counterfactual Account". *Noûs*, 37 (1): 1–24.

Wright, C. (2012). "Mechanistic Explanation without the Ontic Conception." *European Journal for Philosophy of Science*, 2 (3): 375–394.

Wright, C. (2015). "The Ontic Conception of Scientific Explanation." *Studies in History and Philosophy of Science Part A*, 54: 20–30.

Wright, C., and Van Eck, D. (2018). "Ontic Explanation Is Either Ontic or Explanatory, But Not Both." *Ergo*, 5 (38): 997–1029.

Wu, C. H., and McLarty, J. W. (2000). *Neural Networks and Genome Informatics* (Vol. 1). Oxford, UK: Elsevier Science.

Index